General editor:
Brian P. FitzGerald

Science in Geography 4

Data use and interpretation

Patrick McCullagh

Oxford University Press 1974

Oxford University Press, Ely House, London W1

Glasgow New York Toronto Melbourne Wellington
Cape Town Ibadan Nairobi Dar es Salaam Lusaka Addis Ababa
Delhi Bombay Calcutta Madras Karachi Lahore Dacca
Kuala Lumpur Singapore Hong Kong Tokyo

'The aim (of scientific method) is not to open a door to endless knowledge, but to put a limit to continuing error.'—*Brecht*

PRINTED IN GREAT BRITAIN BY OFFSET LITHOGRAPHY BY
BILLING AND SONS LTD., GUILDFORD AND LONDON

Preface

Geography in schools is at present going through a period of change, a change which represents to many a much-needed overhaul, to others an unnecessary dabble in apparently obscure and complicated statistical techniques. Society today is making demands on education which schools and colleges must recognize by taking part in certain changes, if their students are to become adults equipped to play their full part in the society of tomorrow. Much that is still taught is of questionable relevance to the student's needs.

Geography is a discipline which has been slow to carry through at school level the changes that have been taking place in universities. Because of this, ill-defined subject groups, such as environmental studies, social studies, or interdisciplinary studies, are tending to supersede it in schools. Greater co-operation between subjects is admittedly necessary, but one must be aware of the possibility of geography as a school discipline disappearing completely. This would be highly undesirable from an educational point of view, but quite deserved while geography continues to provide little intellectual rigour. The title of a subject, a mode of inquiry, or a field of knowledge is perhaps unimportant, but we do need an intellectually stronger core to what we teach and learn if the essence of geography is not to vanish from schools and colleges.

The changes that *are* beginning to be introduced are making geography more relevant to the needs of students as they become more involved in urban studies and planning, as they begin to analyse the problems of the developing countries, and as they begin to appreciate the problems of resource conservation. These changes are being accepted and becoming established in schools, but there is still very little said about the nature and philosophy of geography on which these changes depend. It is strongly felt that the sixth-form and college student should understand the more important arguments in this field. Only by involving the student in these issues can we justify what is being studied at Advanced Level and beyond.

Preface

The literary, descriptive approach to geography, where geography is treated as an arts subject, still has a definite role to play, but a clearer understanding of the nature of geography is achieved with the scientific approach. Such an approach requires the study of spatial patterns and overall systems of operation; it requires a greater degree of precision in measurement and description; it requires some estimate of the significance of inferences and conclusions drawn from the relationships being studied; and above all it requires an attempt to set up generalized theory from which predictions can be made.

The important tests of the success of the approach are:

(1) whether the student has a better understanding of the organization of society in a spatial (geographical) sense; and

(2) whether he has therefore developed a greater ability to make reasoned decisions based on his improved understanding.

On the first point, generalized 'models' or structures of the working of reality (which form the basis of scientific geography) aid understanding and act as pegs upon which to hang further ideas, concepts, and factual material. As far as the second point is concerned, a scientific approach to geography increases the ability to act upon evidence, and, through the development of general theory, allows decisions to be made which are based on a better understanding of reality. Thus courses of action can be better planned, and a more worthwhile contribution to society can be made.

The four books in the Science in Geography series are:

Developments in Geographical Method by Brian P. FitzGerald
Data Collection by Richard Daugherty
Data Description and Presentation by Peter Davis
Data Use and Interpretation by Patrick McCullagh.

The plan for the series came from an idea of Peter Bryan, from Cambridgeshire High School for Boys, whose advice during all the various stages of producing the books has been of great assistance.

Stonyhurst, August 1973 Brian P. FitzGerald

Acknowledgement
I am indebted to the Literary Executor of the late Sir Ronald A. Fisher, F.R.S., to Dr. Frank Yates, F.R.S., and to Longman Group Ltd., London, for permission to reprint Tables 3 and 4 from their book *Statistical Tables of Biological, Agricultural and Medical Research.*

Contents

Chapter 1

Introduction

The aim

The last decade has seen a marked increase in the use of statistical methods to test conclusions drawn from the material which the geography student collects as a result of work in the field, including the evaluation of data from census and other reports. Starting in the universities, this new approach is becoming increasingly common in sixth form geography, a fact which has been recognized in new G.C.E. syllabuses. The aim of this book is to teach the use of some simple statistical techniques to those who **have no background of 'A' level mathematics,** so that the **interpretation** of information becomes more rigorous. You will see how important this application of statistical techniques is, when you realize how often an apparent correlation, or significant pattern discovered during the course of field work, when the data is tested, turns out to have an unacceptably high probability of simply being the result of chance.

Inferential statistics

The mathematical assessment of probability seems to have started during the seventeenth century with attempts to forecast results in games of chance, such as throwing dice. Today **techniques which deal with probability are known as inferential statistics.** These are important to the geographer because they can be used to test the reliability of samples, and so much of his work must inevitably be done through sampling. (You cannot question everyone in a city about their shopping habits, even if everyone were prepared to answer your questions.) We normally need answers to certain key questions. For example, how confident can we be that the sample we have taken–people in the shopping centre–really reflects the habits of the city population from which it was drawn? Or, if we have the yield per acre of some crop in one year for two distinct areas, to what extent are the observed differences due to random variations in that particular year, or are they so unlikely to be due to chance that they probably represent real regional differences that require explanation? Or we observe from published statistics an apparent association in Britain between areas where people have low personal incomes and areas where unemployment is relatively high. What is the extent of the apparent correlation between these two facts, and what is the probability that we have observed chance variations? Inferential tests help us to find answers to questions of this kind.

Fig. 1.1 in S.I.G. 1 (*Science in Geography,* Book 1) illustrates how the techniques introduced in this book assist in the analysis of geographical problems.

Measurement

There are four different scales with which we commonly measure our data. One is the **ratio scale**, which is distinguished by having a true zero, using interval measurement. If we buy two *pounds* of sugar we should get exactly *twice as much* as if we bought one pound. If we bought two *kilos* of sugar we should again get exactly *twice as much* as if we bought one kilo. And when the indicator points to zero it is because there is nothing in the pan. Thus ratios remain true irrespective of the unit used, and the zero means 'nothing'. Notice that on the ratio scale we know exactly how much more sugar we have if we buy two pounds instead of one. In other words we have a precise measure of the *interval* between one and two pounds. This is known as interval measurement.

Temperature measured in degrees centigrade is also measured on an **interval scale**. We know exactly *how much* hotter one day is than another. But it is not a ratio scale, because 0 °C does not mean there is no temperature at all, but represents the temperature of melting ice at normal pressure. It is also not a ratio scale because a little thought will show that to say a day with a maximum of 10 °C is twice as hot as a day with a maximum of 5 °C is obvious nonsense.

There are some occasions when we know one thing is greater than another, but we do not know by how much. This may be because the data is only available in this form, or it may be that exact measurement is not meaningful. For example, if a sample of people is asked to put five different resorts in order of preference it would hardly be legitimate to ask them to put an exact numerical value on the difference. It is possible to state: 'I would rather go to A for a holiday than B; and I would rather go to A or B than C'. But to allot a precise percentage to a difference in preference would be spurious accuracy. When it is possible to place data *in order*,that is, when it is possible to allot **rank values,** we are said to have achieved measurement on the **ordinal scale.** Ordinal measurement is important, partly because it is useful with the 'preference' type of sample mentioned, but also because some very useful tests make use of the ordinal scale, and it is easy to convert data on a ratio or interval scale to an ordinal scale by allotting a rank to each value.

Sometimes our information only consists of the number of things in one or more categories. For example, we might have observed a difference in agricultural economy between farming in an area of limestone upland and in a clay vale. Field work could disclose the number of farms in each relying for their income, say, mainly on milk production. We should thus have *classified* the farms according to whether they were on the limestone upland or in the clay vale. In other words our data would be in the form of **categories.** Information in this form is said to be on a **nominal scale.** An example of the use of nominal data is contained in the first part of Chapter 2.

Probability theory, which forms a major part of statistics, can be very complex indeed. This does not mean that the methods devised by the statistician are necessarily difficult to use. Unless we are mathematicians we have to accept the theory without being able to prove it, but the application of many statistical methods is quite simple. We may take an analogy from learning to drive a car. It is perfectly possible to become a satisfactory driver, and to understand the capabilities and limitations of a car, without knowing precisely how the differential gear functions. Similarly, it is possible to use simple statistical techniques without the theoretical knowledge that lies behind them, provided the assumptions and limitations of each are known and understood.

Table 1.1 *Statistical tests: their requirements and their application*

Information	Techniques
1) Data in the form of frequencies 2) Nominal measurement achieved by allocating data to categories	χ^2 test for one sample χ^2 test for k independent samples
	χ^2 contingency coefficient
1) Data in ordinal form 2) Values measured on an interval scale which have been allotted ranks	Mann Whitney U-test for two independent samples
	Wilcoxon test for two related samples
	The runs test
	Kendall's rank correlation coefficient Spearman's rank correlation coefficient
1) Data measured on an interval scale with distribution of background population assumed to be approximately normal	z-scores
	Probability paper
	The standard error of the mean The binomial standard error
	Simple regression a) by using semi-averages b) by the least squares method

Function	Page
To test whether an observed distribution of frequencies in two or more categories is significantly different from the expected distribution.	6 49
A descriptive numerical index to indicate the degree of association between two variables measured on a nominal scale.	88
To test whether two independent samples are derived from the same population.	38
As the U-test, to determine whether samples are derived from the same population, but for paired values.	58
A method of determining whether the number of runs in an ordered sequence may be regarded as random.	55
To provide a mathematical measure of the degree of association between variables having paired values.	77 85
To predict the probability of an occurrence equal to or greater or less than a given value, or of any value within a specified range.	23
To assess whether a given set of values is approximately normally distributed. To assess an approximate mean and standard deviation. To predict approximate probability as for a z-score.	31
To assess confidence limits for samples.	66 69
To provide a graphical expression of the summary of the relationship between two variables having paired values, and to allow a rough approximation of the value of one variable for any given value of the other.	95 97

Chapter 2

Probability

Chi-squared

Comparison between areas of equal size

One of the simplest and most versatile statistical tests is that of chi-squared (written χ^2). Let us take a simple fictitious example to see how the test works. Suppose that we are conducting a land use survey of two adjacent valleys, one object of which is to describe and explain the glacial features encountered. The valleys are similar in area, but differ somewhat in aspect and altitude. In valley A 25 separate drumlins are counted, while in valley B there are only 11. It is tempting to conclude that, because there are more than twice as many drumlins in one valley as in another *of the same area,* this must have resulted from a difference in glacial history, caused perhaps by differences in altitude and aspect. We may be right in thinking this to be so, but before we can begin to search for factors which may have caused the unequal distribution of drumlins between the two areas we must first estimate the extent to which the distributions we have observed may have come about simply through 'chance' (see p. 27). To do this, let us represent the two valleys diagrammatically (Fig. 2.1).

We are not concerned here with the precise location of the drumlins; simply that there are more in one valley than in the other. Suppose that in each valley the conditions for the formation of drumlins had been similar, then it is reasonable to expect each valley to contain approximately the same number of drumlins. As the two valleys contain a total of 36 drumlins, in the case of our fictitious example this would mean about 18 drumlins in A, and about 18 in B. Chance variations could, of course, account for differences. What we must do is estimate the probability that the differences we *actually observe* in the field are due to chance. If the probability is very low then we may with confidence seek an explanation. If, however, our estimate indicates a considerable probability that our observed variations may simply be due to chance, there is no justification in looking for causes.

Fig. 2.1 *Diagrammatic representation of the distribution of drumlins in two valleys*

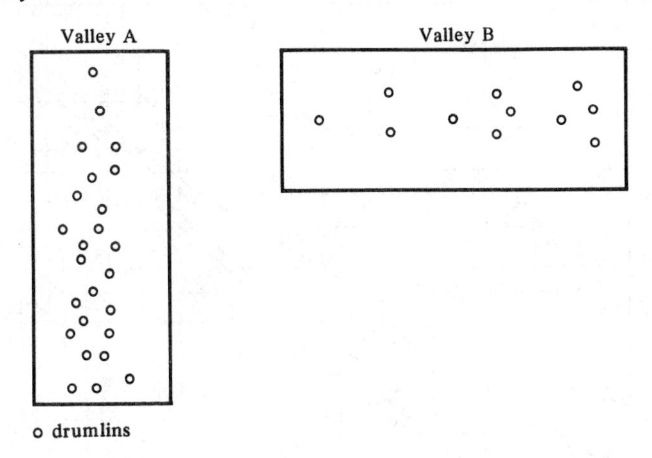

The χ^2 test has been devised for just such a case. It allows us to test an observed distribution against some other hypothetical distribution. In this example the hypothetical distribution is that the drumlins are evenly distributed between the two valleys. We are able to assess the extent to which the difference may be due to chance, by using the formula,

$$\chi^2 = \sum \frac{(O - E)^2}{E}$$

Formula 2·1

where O is the frequency observed, and E the frequency expected (i.e. the frequency that results from the hypothetical distribution, in this case, 18 drumlins in each valley). Σ (big sigma) is a symbol that directs us to **sum** (add up) the results of working out each fraction.

In the case of valley A we have observed 25 drumlins and O therefore is equal to 25. In valley B we observed 11 drumlins, therefore O equals 11. We 'expect' 18 drumlins (25 + 11 evenly distributed between the two valleys), and E is therefore equal to 18. We can now use Formula 2.1;

$$\chi^2 = \sum \frac{(O - E)^2}{E} = \frac{(25 - 18)^2}{18} + \frac{(11 - 18)^2}{18}$$

$$= \frac{49}{18} + \frac{49}{18} = 5{\cdot}4$$

Fig. 2.2 *Probability graph for critical values of* χ^2

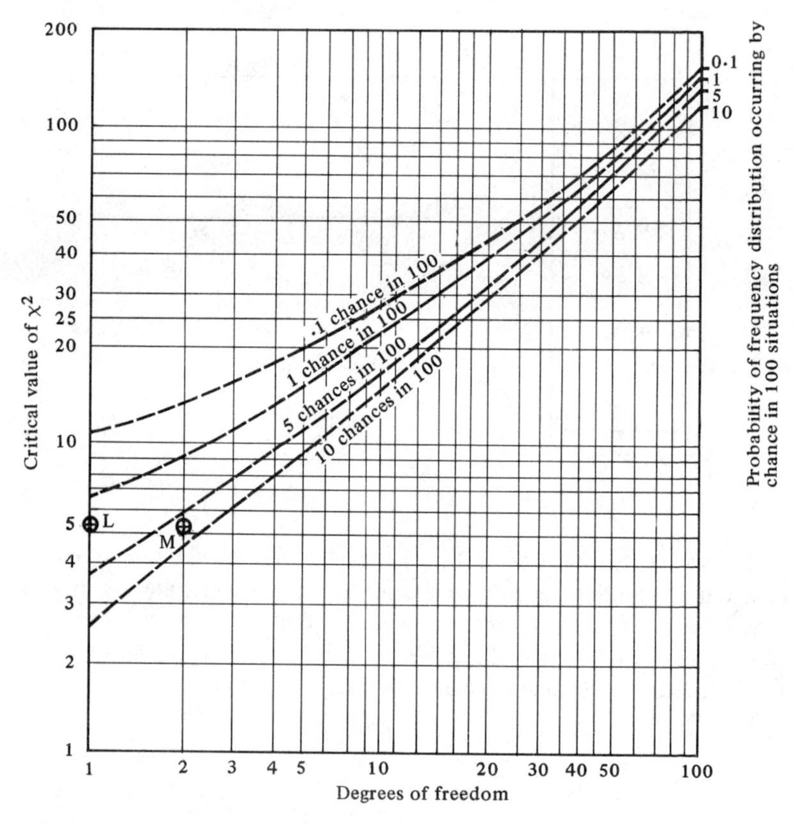

We now have a calculated value of $\chi^2 = 5{\cdot}4$, and are in a position to use Fig. 2.2 to determine the probability that the distribution we are investigating may be due to chance. It will be seen that the horizontal axis of the graph is marked in **degrees of freedom**. The theory that lies behind this need not concern us. For the purpose of the χ^2 test degrees of freedom (usually referred to as *df*) are **the total number of fractions minus one.** In our example we have two fractions, therefore $df = 2 - 1 = 1$.

It is now possible to use the two coordinates we have calculated ($\chi^2 = 5{\cdot}4$, and $df = 1$) to establish a location on the graph. In Fig. 2.2 this location is marked **L**. It will be seen that **L** falls between the pecked (broken) lines 1 and 5 drawn from the scale at the right hand side of the graph. The scale shows the number of times in 100 situations a particular distribution may be expected to occur solely through chance. Because **L** lies between lines 1 and 5, we can conclude that the actual distribution of the drumlins in valleys A and B might be expected mathematically to occur solely through 'chance' variations *more than* once but less than 5 times in 100 occasions. To put this another way, we can be more than 95% sure that the distribution of drumlins between the two valleys is *not* due to chance. Having established this, it is now possible to proceed from a secure statistical base to search for causes as to why the differences exist.

Comparison between areas of unequal size

For the sake of simplicity, in the above example of the use of χ^2 two valleys only were considered and these were assumed to be equal in area. In practice different places under study are very rarely equal in area, and it is often desired to test a distribution over more than two. Such a situation might arise in the study of the frequency of village settlements in relation to geological differences. For example, when villages in midland England were being established between the seventh and thirteenth centuries, the ease with which soil could be worked with primitive implements, the availability of water, and the natural richness of the land were all factors which depended largely on geological characteristics and which tended to make some areas more attractive than others to the early settlers.

The situation in Fig. 2.3 shows the location of 42 villages which, from the evidence of their place names, appear to be early settlements, on three geological outcrops of grit, limestone, and shale. The location of the villages seems to indicate that the limestone particularly was more attractive to the early settlers than the shale and the grit. But once again, the problem is to determine the extent of the probability that the distribution of villages might be largely a matter of chance.

If all three areas shown in Fig. 2.3 had been equally attractive to the early settlers then it would be reasonable to assume that we should find approximately the same number of settlements in each. In other words the distribution of villages would be roughly evenly spread over the landscape, and differences would be negligible, resulting only from such factors as the whims of individuals, or small local differences in topography. This is a case in which we may again use χ^2, although this time we have to modify the technique slightly because the sample areas we have chosen differ in size.

First we have to determine our observed frequencies; in this case the number of villages located on each rock type. Fig. 2.3 shows 20 villages sited on limestone, 16 on shale, and 6 on grit. The problem is to calculate the 'expected frequency' appropriate to each. For this it is first necessary to approximate the *area* occupied by each rock type as a proportion of the total area under investigation (this is easily done by super-imposing a grid), and we find about 32% is limestone, 43% shale, and 25% grit. Now, if each of the three areas was equally attractive to the early settlers, we should expect to find a roughly even distribution of 42 villages over the landscape. It is reasonable to assume in this example that, as limestone occupies 32% of the area, roughly 32% of all the villages (i.e. just over 13) would be found on limestone. Similarly we should expect 43% of the villages to be on shale, and 25% on grit. In other words we calculate the expected frequency for each rock type by allotting the same proportion of the total observed villages as the rock occupies of the total area. This is shown in Table 2.1.

Looking at Table 2.1 we see that we expect 10·5 villages on the gritstone, when in fact there are only 6. On the limestone we expect 13·4 whereas we actually count 20. On the shale the difference between what we expect (18·1) and what we count (16) is rather less. The table thus reveals considerable differences between the kind of distribution which could be anticipated if all areas were equally attractive for settlement, and what we actually observe. It might appear that we should be justified in arguing that

Fig. 2.3 *Distribution of villages between areas of grit, limestone, and shale*

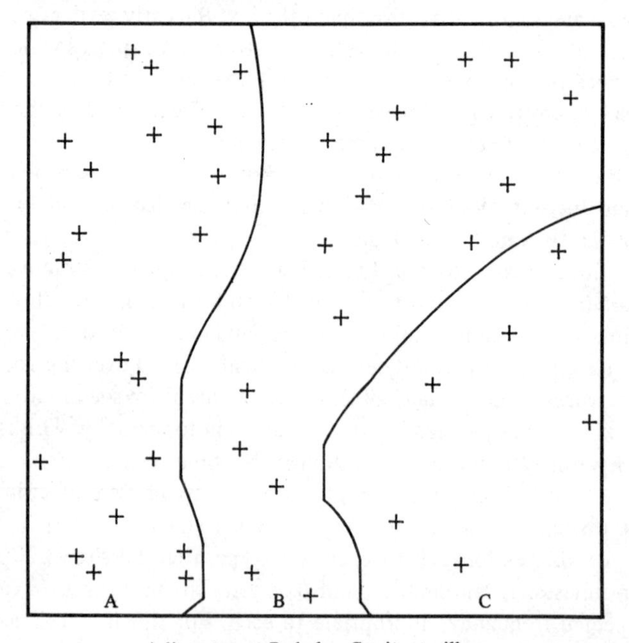

the limestone was specially attractive to settlers, the gritstone much less so, and that this accounts for the different densities of village settlement. The point is, are the numbers of villages on the three types of rock so different that the probability of observed variations being due to chance is acceptably low? The only way of estimating this is by the use of a statistical test. The data is on a nominal scale (the villages having been divided into categories by type of rock), and χ^2 is therefore an appropriate test.

Table 2.1 *Tabulated data of information derived from Fig. 2.3*

Type of rock	Observed frequency of villages	Percentage of total area	Expected frequency
	O		E
Grit	6	25%	10·5
Limestone	20	32%	13·4
Shale	16	43%	18·1
	42	100	42·0

It is now possible to calculate a χ^2 value for the distribution of villages on the three types of rock:

$$\chi^2 = \sum \frac{(O-E)^2}{E}$$

$$= \frac{(6-10{\cdot}5)^2}{10{\cdot}5} + \frac{(20-13{\cdot}4)^2}{13{\cdot}4} + \frac{(16-18{\cdot}1)^2}{18{\cdot}1}$$

$$= 5{\cdot}4$$

Degrees of freedom are the total number of fractions less one. Here $df = 3 - 1 = 2$. Reference to Fig. 2.2 shows that with the χ^2 value again equal to 5·4, but this time with two degrees of freedom, the location on the graph where these coordinates intersect lies at point **M** between the pecked lines 5 and 10. This indicates that (mathematically speaking) a distribution like the one we are testing might occur through chance variations alone more than 5 times in 100. Generally speaking, in a geographical problem a probability as high as this that a distribution might be due solely to chance means that any conclusions concerning causation have an insecure foundation. If research is to proceed on the question of whether rock type affects the number of villages, then a larger sample must be obtained and further tests conducted to see whether there is an acceptably low level of probability that the settlement pattern may be the result of chance.

χ^2 is a simple statistical test to apply, but it is important to note, even though the reasons are too complex to explain here, that there are conditions which must be satisfied if it is to be valid.

1. Data must be of the counted variety. That is to say it must be in the form of **frequencies.** For example, the *number* of any particular features such as drumlins or farms *counted* in a particular area may legitimately be used as an observed frequency, but a percentage number, or numbers per square kilometre, may *not.*
2. Total observed frequencies must equal at least 20.
3. Generally the expected frequency calculated for any fraction should not be less than 5, although it is permissible for 20% of the fractions involved in a calculation to have an expected frequency of less than 5, provided it is not less than 1.
4. A χ^2 test will give you the extent of the probability in *mathematical terms* that a given distribution is due to chance. The interpretation of the result of the test, as with all statistical tests, depends upon the skill and knowledge of the researcher.

The normal distribution curve

Those who have read S.I.G. 3 will understand the use of the **standard deviation,** as one of the most common mathematical techniques to measure dispersion. Equally important, however, are its properties in the estimation of probability. The standard deviation can be used in this way because of its relationship to what is called the **curve of normal distribution.**

Suppose we have the annual rainfall totals for a particular station for 100 years, and that they form the distribution set out in Table 2.2.

Table 2.2 *Annual rainfall totals over 100 years at a fictitious weather station to nearest whole cm*

Measured annual rainfall in cm	Number of years recorded (f)
<60	5
60 – 79	10
80 – 99	20
100 – 119	30
120 – 139	20
140 – 159	10
$\geqslant 160$	5

Note: < 60 is simply the mathematical way of writing 'less than 60'. $\geqslant 160$ means 'equal to or greater than 160'.

It will be noted that the **frequency** (the number) of the years when annual rainfall totals fell within each class interval forms a symmetrical

Fig. 2.4 *A normal distribution*

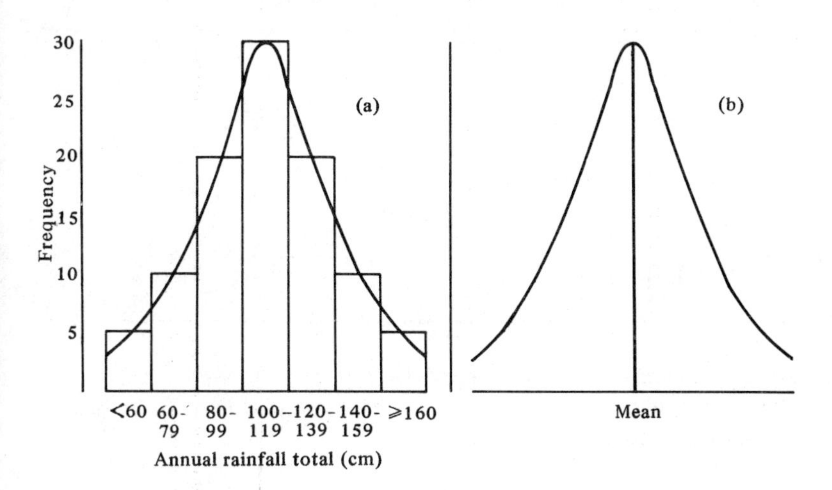

distribution: 5 10 20 30 20 10 5. Rainfall distributions are often shown in the form of a histogram, and this has been done in Fig. 2.4a for the data recorded in Table 2.2. It will be observed that the centre of each bar has been joined by a line to form a smoothed curve. Because the frequency distribution is symmetrical the resulting curve is symmetrical too. Fig. 2.4b shows the same curve without the bars included but with the mean annual rainfall value of 110 cm indicated by the vertical line. Again because the curve is symmetrical, the line indicating the mean divides the area below the curve into two equal parts. This type of symmetrical curve with the mean value bisecting the area enclosed is known as the normal curve, or the curve of normal distribution.

Of course, by joining up the centre point of each bar we have made the assumption that values are distributed evenly within each class interval. This is legitimate in the case of our fictitious example since the object is to show what the 'normal' curve is. In practice, drawing a smooth curve in this way would only be an approximation.

A normal distribution means that all the values (or measurements) concerned are distributed in such a way that if plotted in the form of a frequency histogram, the line joining the centre point of each bar would approximate to the kind of symmetrical curve shown in Fig. 2.4b. A real situation is given in Table 2.3, the annual rainfall totals for 88 years at Chatsworth in Derbyshire from 1878 to 1971 (with six years missing).

Table 2.3 *Annual precipitation at Chatsworth, 1878–1971. Years 1882–4, 1886, 1924, 1961, are missing*

Measured annual precipitation in cm	Number of years recorded
50 – 59	2
60 – 69	8
70 – 79	26
80 – 89	24
90 – 99	13
100 – 109	11
110 – 119	2
120 – 129	2

Fig. 2.5 *Annual precipitation recorded at Chatsworth for 88 years, 1879–1971, with frequencies grouped into class intervals of 10 cm each. Six years are missing*

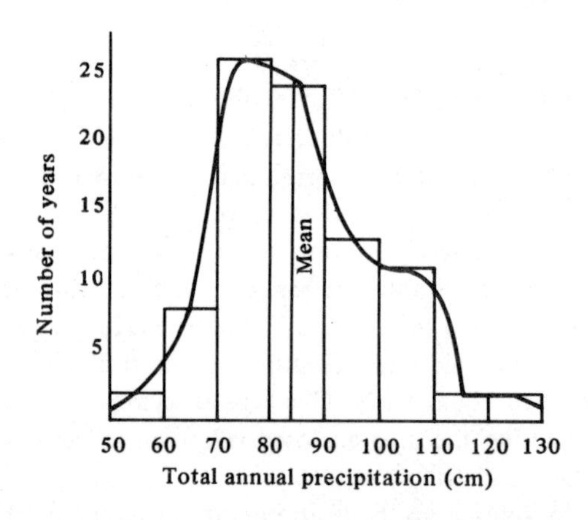

Clearly the frequency distribution in the case of annual precipitation at Chatsworth does not look very symmetrical, and a glance at the histogram in Fig. 2.5 confirms this. Parts of the curve superimposed on the histogram are kinked, and the mean does not quite coincide with the highest point of the curve. This is to be expected because we only have data for 88 years. If it had been possible to know the annual precipitation over several hundred years, then, assuming there is no climatic change, the curve would almost certainly have been a much closer approximation to normal. However, even with only 88 values the distortion is not great, and the mean almost exactly divides the curve into equal areas.

The standard deviation

Standard deviation for individual values

The standard deviation is an exact measure of the **scatter**, or dispersion, of a set of values about their arithmetic mean. It is calculated from the formula:

$$\sigma = \sqrt{\frac{\Sigma(x - \bar{x})^2}{n}}$$

Formula 2.2a

where σ (little sigma) represents the standard deviation; x is each value in the data set; $\bar{x}$ is the mean of all the values in the set; and n is the total number of values. (Note that there is a small modification of this formula, see p. 68, which gives a slightly bigger answer.)

Formula 2.2a tells us to calculate the arithmetic mean of all the values ($\bar{x}$); then work out the variation (or difference) between each value (x) in turn and the mean ($\bar{x}$), giving $(x - \bar{x})$; square the result $(x - \bar{x})^2$; then add all the squares together $\Sigma(x - \bar{x})^2$; divide by the total number of values $\Sigma(x - \bar{x})^2/n$ to obtain the mean of the squares of all the variations and finally obtain the square root of this figure $\sqrt{[\Sigma(x - \bar{x})^2/n]}$. Thus the standard deviation is a measure of the extent to which the individual values of a given set of data are clustered relatively closely around the mean, or whether there are large differences between the mean and some of the values. If there are big differences then large numbers will result when the number is squared. The mean of the squares of all the differences from the sample mean is called the **variance**. It follows that the greater the differences of individual values from the mean the larger will be the variance, and the standard deviation (which is the square root of the variance). In this way it will be seen that the standard deviation is a measure, reduced to a single number, of the extent of all the differences between a set of numbers and their arithmetic mean.

A simple example will show how Formula 2.2a works. Suppose we wish to calculate the standard deviation of the following series of numbers:

1 2 3 4 5 6 7 8 9.

It is convenient to tabulate the data and, having found the mean (in this case 5), calculate the other required values as follows:

x	$(x - \bar{x})$	$(x - \bar{x})^2$
1	−4	16
2	−3	9
3	−2	4
4	−1	1
5	0	0
6	1	1
7	2	4
8	3	9
9	4	16
$\Sigma x = 45$ $n = 9$		$\Sigma(x - \bar{x})^2 = 60$

$$\text{The mean } (\bar{x}) = \frac{\Sigma x}{n} = \frac{45}{9} = 5$$

It is now possible to find the standard deviation by substituting these values in Formula 2.2a.

$$\sigma = \sqrt{\frac{\Sigma(x - \bar{x})^2}{n}}$$

$$= \sqrt{\frac{60}{9}} = 2{\cdot}58$$

This example has been given because it shows in the simplest way how to calculate a standard deviation. Whole numbers (or **integers**) were chosen so that the mean is also a whole number, which greatly simplifies the arithmetic. Very often, however, the mean will not be a whole number, and the resulting decimal will make computation longer and increase the liability to error. Fortunately, an alternative formula is available, derived algebraically from Formula 2.2a (and so yielding exactly the same answer). This alternative formula reduces the work involved and is easier to operate, especially if a simple calculating machine is available.

$$\sigma = \sqrt{\frac{\Sigma x^2}{n} - \left(\frac{\Sigma x}{n}\right)^2}$$

Formula 2.2b

Let us now take two real sets of data and see what the standard deviation will tell us. The figures are the annual precipitation totals for Sutton Coldfield in England, and Canberra, Australia, for the 20 years 1941–60.

Table 2.4 *Annual precipitation in cm, 1941–60*

Sutton Coldfield		Canberra	
x	x^2	x	x^2
80	6400	50	2500
64	4096	65	4225
58	3364	62	3844
61	3721	31	961
56	3136	57	3249
92	8464	57	3249
62	3844	71	5041
81	6561	82	6724
60	3600	70	4900
72	5184	110	12100
86	7396	56	3136
72	5184	96	9216
55	3025	49	2401
83	6889	48	2304
58	3364	78	6084
59	3481	103	10609
69	4761	37	1369
89	7921	77	5929
64	4096	87	7569
94	8836	79	6241
$\Sigma x = 1415$	$\Sigma x^2 = 103\ 323$	$\Sigma x = 1365$	$\Sigma x^2 = 101\ 651$
$n = 20$		$n = 20$	

Using Formula 2.2b, the calculation of the standard deviations for the figures for Sutton Coldfield and Canberra is as follows:

Sutton Coldfield

$$\sigma = \sqrt{\frac{\Sigma x^2}{n} - \left(\frac{\Sigma x}{n}\right)^2}$$

$$= \sqrt{\frac{103\ 323}{20} - \left(\frac{1415}{20}\right)^2}$$

$$= 12{\cdot}7$$

Canberra

$$\sigma = \sqrt{\frac{\Sigma x^2}{n} - \left(\frac{\Sigma x}{n}\right)^2}$$

$$= \sqrt{\frac{101\ 651}{20} - \left(\frac{1365}{20}\right)^2}$$

$$= 20{\cdot}6$$

Fig. 2.6 shows the annual totals plotted in the form of a histogram. The mean annual precipitation for Sutton Coldfield ($\overline{x} = 70{\cdot}7$) is very similar to that of Canberra ($\overline{x} = 68{\cdot}2$). But it will be observed how much greater is the *variability* (i.e. the fluctuations from year to year) in Canberra. The

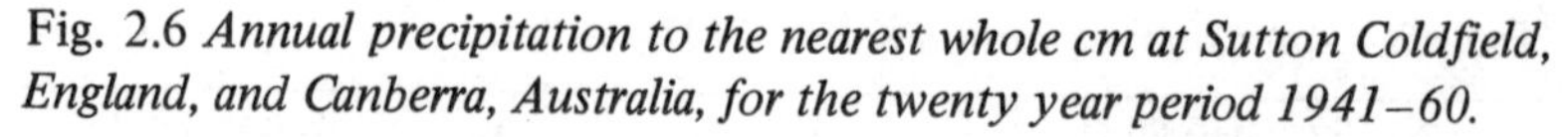

Fig. 2.6 *Annual precipitation to the nearest whole cm at Sutton Coldfield, England, and Canberra, Australia, for the twenty year period 1941–60.*

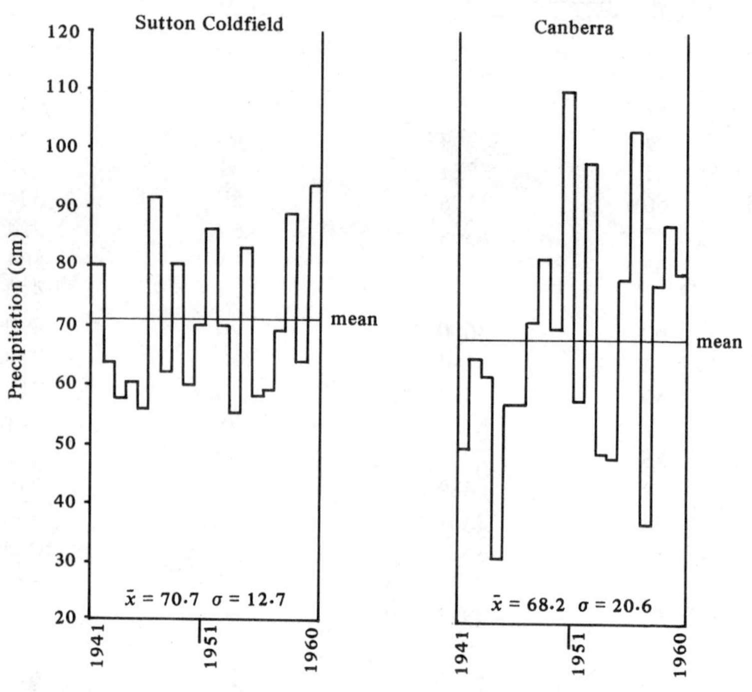

highest and lowest totals for Sutton Coldfield are 94 cm and 55 cm while those for Canberra are 110 cm and 31 cm. This difference in variability is reflected in the standard deviations. Whereas the standard deviation for Sutton Coldfield is 12·7, the standard deviation for Canberra is 20·6, very nearly twice as large. Fig. 2.6 gives a clear visual impression of the difference. The mean annual precipitation total tells us a certain amount, but if we calculate the standard deviation as well we know a great deal more about the rainfall regime.

Probability and the normal curve

We have seen (p. 12) what we mean by 'normally distributed data', and 'the normal curve'. Its importance to us here lies in its relationship to the standard deviation. The mathematical theory of this is not easy, but fortunately, there is no difficulty in understanding the principle, that **if a set of**

data has a near normal distribution, then just over 68% of the values will fall within one standard deviation (plus and minus) of the mean; just over 95% of the values will fall within two standard deviations of the mean; and just over 99% will fall within three standard deviations of the mean.

Fig. 2.7 *Relationship between the normal curve of distribution and the standard deviation of normally distributed data.*
68% of values will lie plus or minus one standard deviation from the mean.
95% of values will lie plus or minus two standard deviations from the mean.
99% of values will lie plus or minus three standard deviations from the mean

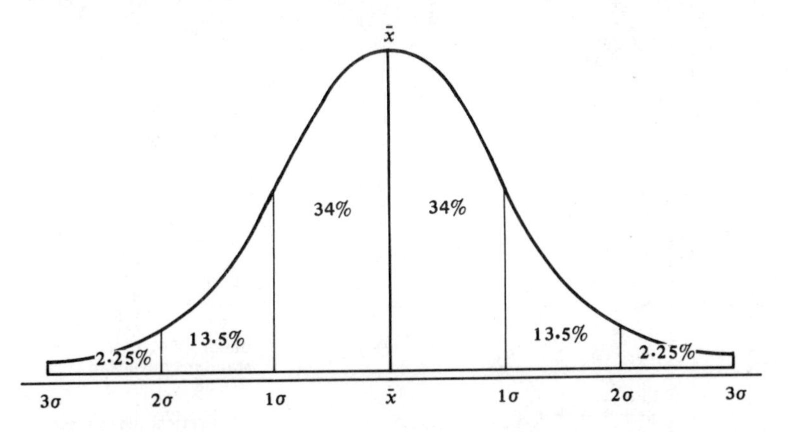

The normal curve of distribution is shown in Fig. 2.7. The percentage of values falling one, two, or three standard deviations from the mean are only true when a large number of values is involved. If the data consists of many thousands of numbers, then the percentage of values falling within one, two, or three standard deviations of the mean will relate very closely indeed to the percentage distribution in Fig. 2.7. On the other hand, if relatively few numbers are involved, for example 100, then the percentage figures for the curve are a close approximation only.

You may have noted that the percentage figures total 99.5%. This means that about 0·5% of the values may be expected to lie more than three standard deviations from the mean, and if we have only one hundred numbers it is not even possible for exactly 0·5% of them to do so.

We can look at the normal distribution in a different way. The curve in Fig. 2.7 is a **probability curve.** The scale on which the actual *value* of each number is measured is along the base. The probability of one or more of those values occurring is given by the *area* enclosed below the curve.

Fig. 2.8 *A dispersion graph showing annual precipitation at Chatsworth, Derbyshire, for 88 years from 1878–1971 (with six years missing). For explanation see text*

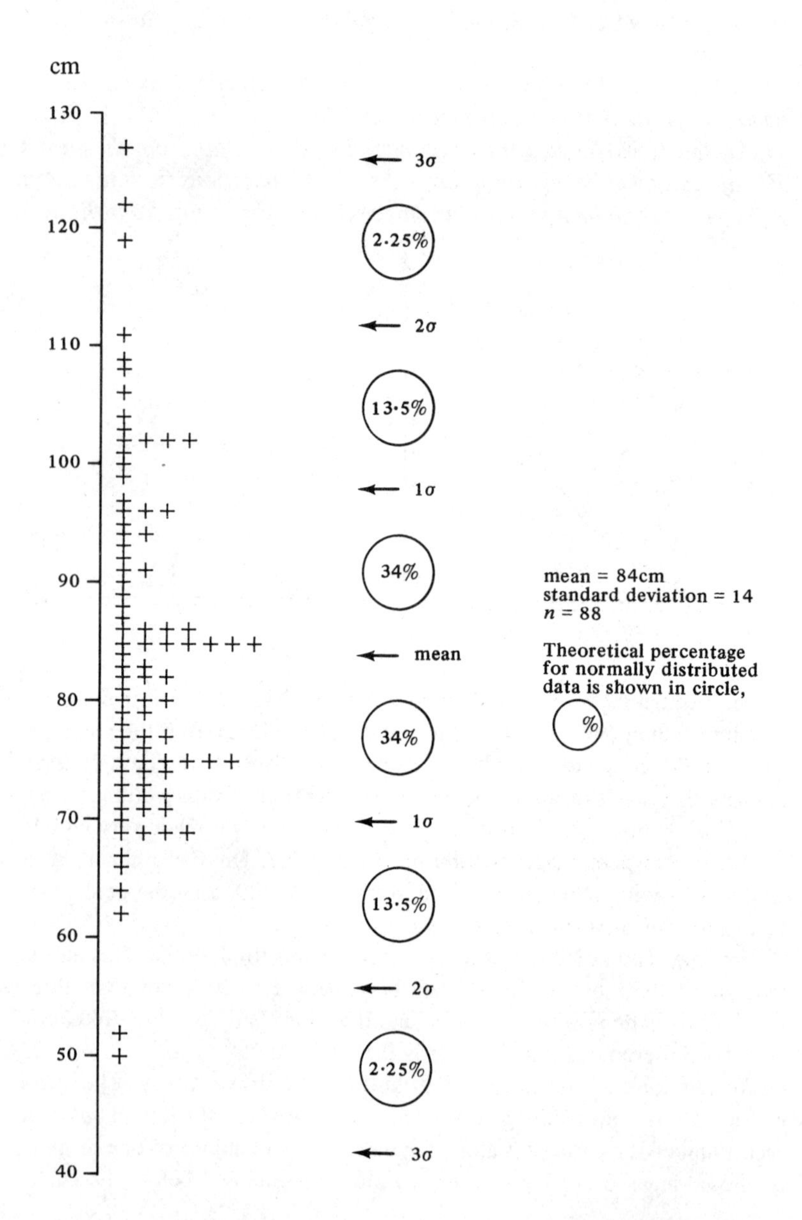

What we *can* say (and this is why the standard deviation is so important) is that there is a 0·5% *probability* that one of our hundred numbers will be *more* than three standard deviations more or less than the mean. In other words if our set of data consisted of thousands of values (theoretically an infinite number) 0·5% of all those values would be more than three standard deviations from the mean.

This concept becomes clearer when demonstrated by an actual example. If we look back to Fig. 2.5 we see the 88 annual precipitation totals recorded at Chatsworth in Derbyshire grouped in intervals of 10 cm. The lowest precipitation recorded was 50 cm in 1921, and the highest 127 cm in 1880. You will see that the distribution is only approximately normal, resulting in the smoothed curve being 'kinked' and rather skew. This is exactly what one would expect when only 88 values are involved. Nevertheless, the actual curve approximates to the theoretical normal curve, and the mean almost exactly bisects the area beneath it. (Note that the values for each year are measured on the scale along the base. It is possible to draw a histogram because the values have been grouped in class intervals.) If we ignore the irregularities of the curve and assume it approximates to normal, then we should expect the distribution of values of annual precipitation to correspond approximately with the percentages shown in Fig. 2.7.

The actual values for annual precipitation at Chatsworth over a period of 88 years are shown in Fig. 2.8 in the form of a dispersion graph. The calculated mean of all the annual totals is 84, and the standard deviation is 14. It will be found that 69·32% of the observed annual values fall within one standard deviation of the mean; 25% are between one and two standard deviations from the mean; and 4·55% are between two and three standard deviations from the mean. This is a very close correspondence indeed with the theoretically expected distribution of the normal curve (Fig. 2.7), and is obtained despite the small number of values involved and the deviations of the observed facts from a perfect normal distribution. One value only lies slightly more than three standard deviations from the mean. This is not remarkable as 0·5% of values (when a large number are involved) might be expected to do so. In this case one value represents 0·88% of all the observations involved. (The percentage figures do not quite add up to 100 because the decimals have been rounded off.)

Based on the observations we have for Chatsworth and the theoretical distribution of values related to the normal curve it is possible to predict that for a period of 100 years:

in 68 years the precipitation will be between 70 cm and 98 cm;
in 95 years the precipitation will be between 56 cm and 112 cm;
in 5 years the precipitation will be below 56 cm or over 112 cm.

Water supply authorities rely on climatic statistics presented in this way when planning new reservoirs. Of course, it is not possible to predict which years will be very wet and which very dry. All that can be forecast is the percentage of years in which the annual precipitation will be between certain values. From this the effect on the water supply can be estimated.

Standard deviation for grouped frequencies

When the set of data is very large it is convenient to group the values into class intervals. (Some published data, e.g., of the size of land holdings, is very often only available in this form.) In this case, it is possible to calculate a standard deviation for group frequencies by using a modified formula. However, it should be clearly understood that by placing data in class intervals some information is inevitably lost, because we no longer know precisely where within the class interval a particular value falls. For the purpose of the calculation we have to allot all the values within one class the value of the class mid-point, irrespective of where individual values lie. This introduces an error which must be acknowledged and accepted if the technique is to be used. In practice the difference is generally relatively small because errors within class intervals tend to compensate each other, especially if a very large number of values is being used.

The formula for calculating the standard deviation of a grouped frequency distribution is:

$$\sigma = c \sqrt{\frac{\Sigma d^2 f}{\Sigma f} - \left(\frac{\Sigma df}{\Sigma f}\right)^2}$$

Formula 2.3

where c is the class interval; f is the frequency within each class; and d is the number of class intervals above (+) or below (–) the assumed mean.

To use Formula 2.3 first decide by inspection which class mid-point is nearest to the mean of the whole distribution. This is the **assumed mean.** It is purely a guess, and will not affect the final answer. If the guess is badly out it will simply result in large, rather unwieldy numbers in the calculation, and therefore more work and chance of error.

Let us apply Formula 2.3 to the Chatsworth annual precipitation totals set out in Table 2.3. In this case we have taken the assumed mean as 74·5cm.

Precipitation (to nearest cm)	Class mid-point	f	d	df	d^2f
50 – 59	54·5	2	−2	−4	8
60 – 69	64·5	8	−1	−8	8
70 – 79	74·5	26	0	0	0
80 – 89	84·5	24	1	24	24
90 – 99	94·5	13	2	26	52
100 – 109	104·5	11	3	33	99
110 – 119	114·5	2	4	8	32
120 – 129	124·5	2	5	10	50
		$\Sigma f = 88$		$\Sigma df = 89$	$\Sigma d^2f = 273$

$$\sigma = c\sqrt{\frac{\Sigma d^2 f}{\Sigma f} - \left(\frac{\Sigma df}{\Sigma f}\right)^2}$$

$$= 10 \times \sqrt{\frac{273}{88} - \left(\frac{89}{88}\right)^2}$$

$$= 14{\cdot}5$$

The standard deviation for these figures using individual values is 14·3, correct to the first decimal place. A standard deviation of 14·5 using grouped data is thus a very close approximation. The difference between the two (0·2) is a result of the information lost due to compressing individual observations to a class mid-point value.

z-scores

A further use of the standard deviation is in the calculation of a z-score (pronounced 'zee'). Suppose, for example, that we wish to know the probability of precipitation at Chatsworth being more than 105 cm in any one year. 105 cm lies more than one and less than two standard deviations from the mean. What we require are the probabilities associated with values in between whole standard deviations. This is done by expressing any specific value in terms of the number of standard deviations it is from the mean (which is a z-score) using the formula:

$$z = \frac{x - \bar{x}}{\sigma}$$ *Formula 2.4*

where x is the specific value, $\bar{x}$ the arithmetic mean, and σ the standard deviation of all the values.

Since we know the Chatsworth mean is 84, and the standard deviation is 14·5,

$$z = \frac{105 - 84}{14{\cdot}5} = 1{\cdot}5 \quad \text{(approximately)}$$

In other words 105 cm is about 1·5 standard deviations above the mean. Appendix 1 is the normal distribution table showing the probabilities associated with various values of z. For $z = 1{\cdot}5$ we see that the probability is $\frac{\cdot 134}{2} = \cdot 067$ (or 6·7%). So we might expect precipitation at Chatsworth to be 105 cm or *more* in six or seven years out of every hundred.

Similarly a z-score of about 0 indicates a value very close to the mean; in the case of Chatsworth we know this is 84 cm. Thus there is a 50% chance of having in any one year rainfall greater than 84 cm, and a 50% chance of having less than 84 cm.

The probability that Chatsworth will receive precipitation *between* any two given values may be determined first by calculating a z-score for each. Then we find the probability associated with each z-score from Appendix 1. The area of the normal curve falling between these two scores represents the probability that the annual precipitation will be *between* the two required values. An example will make this method clear.

Let us suppose we wish to find the probability of annual precipitation at Chatsworth falling between 75 cm and 100 cm. First we calculate the z-scores:

$$\text{For 75 cm,} \quad z = \frac{75 - 84}{14{\cdot}5} = \frac{-9}{14{\cdot}5} = -0{\cdot}62$$

$$\text{For 100 cm,} \quad z = \frac{100 - 84}{14{\cdot}5} = \frac{16}{14{\cdot}5} = 1{\cdot}10$$

From Appendix 1 (Column A) we see that the probability of the precipitation being as little as 9 cm below the mean is ·226. Similarly we find the probability of its being as high as 16 cm above the mean is ·364.

Fig. 2.9 *Probability that precipitation at Chatsworth will fall between 75 cm and 100 cm. For explanation see text*

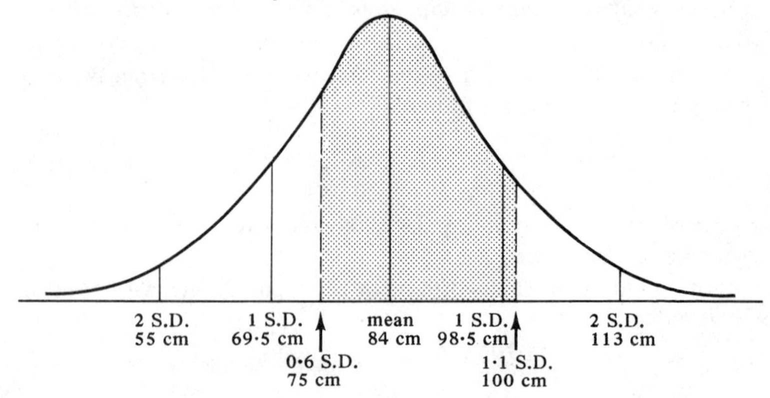

Therefore the probability that precipitation will be as extreme as these two values, *or will fall between them,* is ·226 + ·364 = ·590. In other words, one might expect precipitation at Chatsworth to be between 75 and 100 cm in 59 years out of 100. Fig. 2.9 shows how these particular z-scores relate to the probabilities associated with the normal curve.

Exercises

1. Two rural areas of equal size are randomly selected, one on Upper Red Sandstone, and the other on Bunter Sandstone. The number of occupied dwelling houses on each is counted, and 45 are found on the Upper Red Sandstone, but only 15 on the Bunter Sandstone.

What is the probability that this distribution is due to chance?

2. In an isolated area of Amazonia you discover three tribes of Indians, each having about the same number of people in it. Tribe A lives on the flood plain of a river. Tribe B lives near the river but on higher land. Tribe C lives high on the slope of a mountain close by. During the year you are there 35 people from Tribe A die from disease. The corresponding figures for Tribes B and C are 23 and 17. What is the probability that the differences in mortality rates are due to chance?

3. A survey of abandoned agricultural holdings is carried out in an area of Highland Britain, with the object of relating this to aspect. The data collected are as follows:

Area of Sample	6 km^2	4 km^2	5 km^2	5 km^2
Abandoned Holdings	16	12	5	3
Aspect	Northerly	Easterly	Westerly	Southerly

Assuming that aspect is the major factor in the decision to abandon land, what can be said about the data collected?

4. At Chatsworth the annual mean precipitation is 84 cm, with a standard deviation of approximately 14. What is the probability that, in any one year, precipitation will be:

(i) less than 84 cm;
(ii) less than 70 cm;
(iii) more than 98 cm;
(iv) more than 105 cm;
(v) less than 60 cm;
(vi) more than 125 cm;
(vii) between 70 and 98 cm;
(viii) between 75 and 105 cm;
(ix) between 75 and 90 cm;
(x) between 110 and 120 cm.

5. The following figures (given to the nearest whole cm for ease of calculation) are the annual precipitation totals at Cropston near Leicester, from 1872–1966. Years 1891 and 1954 are missing.

1872	92	1890	57	1910	79	1930	96	1950	66
3	59	2	57	1	59	1	81	1	80
4	51	3	56	2	96	2	74	2	69
5	91	4	60	3	66	3	59	3	58
6	83	5	64	4	73	4	62	5	65
7	73	6	61	5	88	5	80	6	68
8	74	7	65	6	77	6	76	7	71
9	73	8	61	7	69	7	79	8	81
1880	90	9	70	8	66	8	64	9	54
1	74	1900	78	9	82	9	81	1960	93
2	95	1	63	1920	73	1940	73	1	61
3	80	2	60	1	56	1	74	2	59
4	52	3	83	2	88	2	59	3	66
5	74	4	66	3	80	3	55	4	53
6	81	5	57	4	87	4	65	5	82
7	48	6	70	5	71	5	57	6	81
8	62	7	71	6	76	6	82		
9	72	8	64	7	95	7	55		
		9	69	8	85	8	78		
				9	72	9	71		

From these figures:

(a) Construct a frequency histogram similar to that in Fig. 2.5, using a class interval of 2.

(b) Calculate the standard deviation of this distribution using Formula 2.3.

Chapter 3

Hypothesis testing

Randomness and chance

The nature of the geographer's data and the use of samples were discussed in S.I.G.s 2 and 3. It is now necessary to consider how these data can be tested, before conclusions may properly be drawn from them. The geographer is primarily concerned with the spatial aspects of problems, and with the consequences of location. He tries to establish methods of comparison between places and areas, and to search for causal relationships between areally distributed phenomena, i.e. the causes and consequences of location. He seeks explanations for observed facts, and attempts to formulate theories. In all these activities he must first eliminate the possibility that what he is observing is simply the result of chance. If it is shown that a distribution is likely to be due to chance (for example, a specific pattern of rural settlement), then obviously there is little point in attempting to explain the reasons for this particular distribution. One of the reasons for the increasing use of statistical techniques in geography is to estimate the probability that observed facts are the result of chance.

Probability (in the sense used above) is dealt with in the next section, but what do we mean by 'chance'? We have already seen (Chapter 1) how χ^2 test can be used to determine the number of times in 100 occasions that a particular distribution is likely to occur (mathematically speaking) by chance. Here chance may be seen in terms of the probability of any particular number between 1 and 6, say a 4, coming uppermost when a fair die (that is a die perfectly balanced and symmetrical in every way) is thrown by an unbiased shaker. It is obvious that every side has an equal opportunity, but that a number of unpredictable facts have caused the die to come to rest with the 4 uppermost. Small differences such as minute irregularities in the shaker, or in the surface on to which the die is cast, will make each throw slightly (but undetectably) different. Everything has a cause, and it is through such a chain of unascertainable facts that the die comes to rest at last. We cannot therefore predict any individual throw, we can only say that each side has an equal probability of coming up, and that whichever one does so is the result of 'chance'.

Of course, if very many throws of the die were made, say 60 000 under identical conditions, we could say with confidence that there would be almost exactly 10 000 1s, 10 000 2s, and so on, because on every occasion the probability that any one of the six faces would be uppermost is equal (i.e. 1 in 6 or 1/6).

Similarly, when we are concerned with geographical problems and speak of 'chance', we mean those unidentifiable facts not directly connected with our problems, but which may affect the result in an apparently random fashion. The storm of unusual intensity that affects a crop yield, the unexpected death of a landowner that brings property on to the market, the influence on the location of a factory of the liking of the general manager for fly fishing; things of this kind may be regarded as 'chance' factors.

The term **random** is used in much the same way. Suppose we have a piece of paper with a grid drawn upon it with the vertical lines numbered 00 to 49, and the horizontal lines numbered 50 to 99. And suppose we had a bag with many thousands of digits 0 to 9 each on separate pieces of card. Provided the bag was properly shaken to mix the numbers thoroughly, and we drew out the pieces of card in pairs without looking, the numbers we drew each time would be the result of chance, and therefore random (i.e., in no ascertainable order). If these numbers were then used to call coordinates for the grid and each location marked, the pattern that would result would be random, because the location of every point would be independent of any other point and have an equal chance of being located on any intersection.

It is in this sense that 'chance' and 'random' are used. **This book is primarily concerned with techniques that establish the extent of the probability that a given situation, when expressed in numerical terms, is due to 'chance' factors.**

The null hypothesis

The use of the hypothesis in geography was explained in S.I.G. 1, Chapter 1. Its application in statistical analysis must now be discussed. To the ancient Greek philosophers a hypothesis was a foundation upon which argument or explanation could be based. In statistics the meaning is almost exactly the reverse: a hypothesis is something which needs to be tested. We may use statistical tests to compare two (or more) sets of different data to try to establish whether there exists a mathematical relationship between them (**correlation**) or whether we may regard two samples as coming from the same population. (A **population** in the statistical sense refers to the total number of values of the distribution with which we are concerned. If we were studying wheat yield per hectare in East Anglia from a sample of twenty parishes, then we might regard the population as the values of wheat yield per hectare for all the parishes in the area.) Thus if we had two samples of wheat yield per hectare from two contrasting areas, A and B, we could use a statistical test to establish whether the two sets of values were sufficiently similar to be likely to form part of the same distribution (or population), or whether they were so different as to be unlikely to form part of one distribution (or population). In other words we could establish whether observed differences in yield had a high probability of being the result of chance variations, or whether the variations were unlikely to be the result of chance and therefore represented a real regional difference in agriculture between A and B.

In order that a statistical test may be used it is necessary first of all to set out the proposition in precise terms. It is customary to do this by the formulation of two hypotheses, written symbolically as H_0 and H_1. H_0 is **the null hypothesis.** It is termed 'null' because it states that the two samples form part of the same population, and there is a high probability that the observed differences are due to chance variations. H_1 is the alternative hypothesis and states that the observed differences are so great that they are unlikely to be the result of chance, and the two samples must therefore be regarded as coming from different populations.

If we consider our hypotheses for wheat yield per hectare in areas A and B, H_0 will state that 'there is no difference between A and B in terms of wheat yield per acre'. H_1 will state that 'the observed differences between A and B are so great it is unlikely that the two sets of values form part of the same population'. Only if the result of the test leads us to reject H_0 and accept H_1, may we confidently begin to look for *reasons* why this difference should exist.

Significance and rejection levels

So far I have spoken rather loosely of differences 'likely to be the result of chance'. In statistics **significance** has a precise meaning, and to understand this we must return to the normal curve. We have already seen (p. 19) the relationship between the standard deviation and the curve of normal distribution, and the approximate percentage of values lying between one,

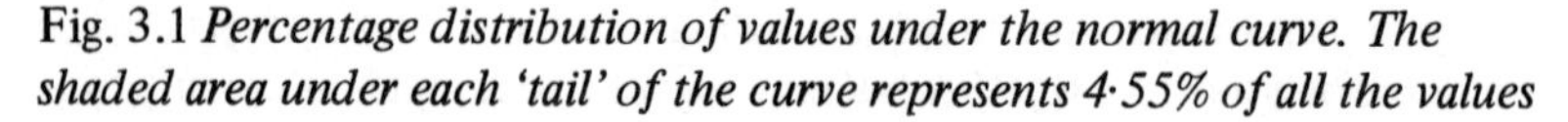

Fig. 3.1 *Percentage distribution of values under the normal curve. The shaded area under each 'tail' of the curve represents 4·55% of all the values*

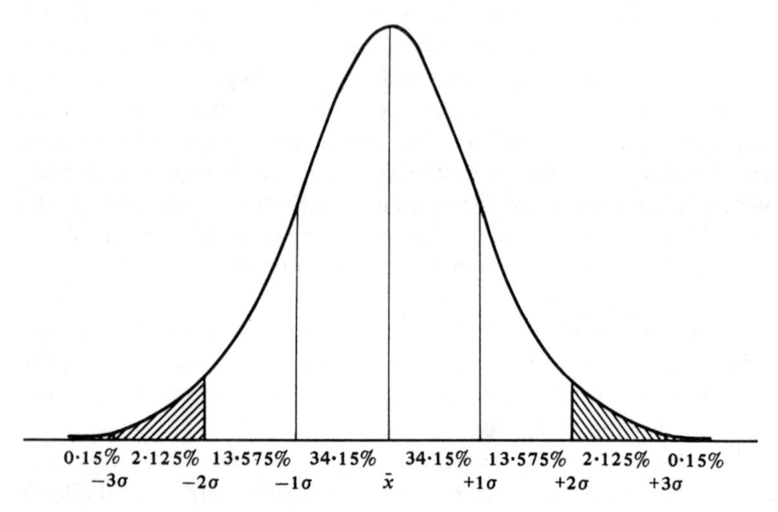

two, and three standard deviations from the mean. Fig. 3.1 shows a similar curve, but the percentages are given in more exact terms. It will be seen that 95·45% of the values are within two standard deviations plus and minus of the mean. It follows that if we select one value *randomly* from the distribution the probability that it would lie *more than two standard deviations from the mean is* 4·55%.

Let us return to our samples of wheat yields. We have one sample from East Anglia. Now if we take very many samples (in theory an infinite number) from a given population it can be shown that **the sample means form a normal distribution.** This is known as the sampling distribution, and the values of the sample means will fall within the percentages shown in Fig. 3.1. Every statistic used in the tests described in this book has its own sampling distribution, and generally when the sample is sufficiently large this distribution will approximate to normal. It is thus possible to predict the probability of occurrence of any particular value, where large samples are involved, by the use of a *z*-score. If the value of the *z*-score is two standard deviations (or more) from the mean, then the probability of occurrence is 4·55% (or less). Therefore, if we are calculating the probability that two samples are drawn from the same population and obtain a result of 4·55% we are able to state with more than 95·45% confidence that the samples are *not* drawn from the same population. If the probability turns out to be only 0·3% (i.e. a *z*-score of three standard deviations) then we can be sure with 99·7% confidence that the two samples come from distributions so different that they must relate to different populations. In practice the decimals are usually ignored, and we speak of 'with 95% confidence, or 'with 99% confidence'.

But probability is frequently referred to on a scale from 0 (will never occur) to 1 (is certain to occur), and 95% confidence is often called **the ·05 level of significance**, meaning that the particular result has only a 5% chance of occurring through random variations. Similarly 99% confidence is referred to as the ·01 level of significance. That is, the result will happen by chance less than once in a hundred times. Associations of numbers that are likely to occur through chance variations less than five times in a hundred are said to be **statistically significant.**

Generally in geographical problems the ·05 level of significance is regarded as sufficiently rigorous. But the actual percentage of confidence (i.e. rigour) that is required for any particular problem can only be decided by the researcher. This degree of confidence is known as the **rejection level** (i.e. the level of significance at which we decide H_0 may safely be rejected), and represented symbolically by α.

Thus if we decide that the ·05 level of significance is sufficiently rigorous in the case of our wheat yield samples from areas A and B (i.e. $\alpha = {\cdot}05$), and if the result of the test shows that there is at least a 95% probability that the values in the two samples are not from the same population, we can reject H_0 and accept the alternative hypothesis, H_1. We now have a sound statistical basis from which to seek reasons for the difference in yield.

In the case of small samples the sampling distribution will not necessarily approximate to normal and a *z*-score cannot be used. (The size of 'small' depends on the statistical test. It varies from about 10 to 30 values in the sample, and is defined for each test. **Over 30 values may always be regarded as a large sample.**) It is possible, however, to calculate the sampling distributions for small samples and to prepare tables from which the significance of the result may be obtained. Tables for each test are given in the appendices.

Note: In the calculation of *z*-scores that follow, no account is taken whether *z* is likely to be plus or minus. The probability associated with any calculated value of *z* is thus the value given in Appendix 1, Column B. This is known to statisticians as a 'two tailed' probability (i.e. a probability including both 'tails' of the normal curve). More advanced texts also include the consideration of 'one tailed' probabilities. These are not considered here because the prediction of whether a calculated z-score is likely to be plus or minus requires experience and considerable geographical knowledge. The probability which results from any calculated value of z for the inferential tests in Chapter 5 should thus always be that given in Column B of Appendix 1.

Table 3.1 *Precipitation at Chatsworth grouped in class intervals of 5 cm*

Precipitation in cm	Frequency	Cumulative frequency	Cumulative percentage
50 – 54	2	2	2·3
55 – 59	0	2	2·3
60 – 64	2	4	4·6
65 – 69	6	10	11·4
70 – 74	9	19	21·6
75 – 79	17	36	40·9
80 – 84	9	45	51·1
85 – 89	15	60	68·2
90 – 94	5	65	73·9
95 – 99	8	73	83·0
100 – 104	7	80	91·0
105 – 109	4	84	95·6
110 – 114	1	85	96·7
115 – 119	1	86	97·8
120 – 124	1	87	98·9
125 – 129	1	88	100·0

Probability paper

The figures for precipitation at Chatsworth are again reproduced in Table 3.1, but this time the annual totals have been grouped in class intervals of 5 cm, and the frequency (i.e. the number of years with values within this range) inserted for each group. The cumulative frequency for each interval has been calculated as a percentage of the total frequency (88 years) and summed to form the cumulative percentage frequency column. (A 'cumulative frequency', or 'cumulative percentage' is arrived at by adding each value in turn to the sum of the values that have preceded it.) The cumulative percentage frequency curve for these figures is given in Fig. 3.2a.

It will be seen from Fig. 3.2a that the cumulative percentage frequency curve (like the ordinary cumulative frequency curve) forms the shape of a rather flattened S, which is typical of this type of curve plotted on ordinary graph paper. There is, however, another type of graph paper also having the vertical scale in terms of percentages, but adjusted so that the intervals close to 50 per cent are compressed, and those approaching 99·99 per cent and 0·01 per cent are greatly extended. Such paper is known as probability paper, because on it a series of values that are perfectly normally distributed will form a straight line. Fig. 3.2b shows the cumulative percentage figures from Table 3.1 plotted on probability paper. They do not form a perfectly straight line because their distribution is not absolutely normal. Nevertheless, the distribution is definitely linear and approximates as closely to a straight line as one usually finds when employing real data.

A best fit line has been inserted roughly by eye through the points of the cumulative percentage values, and this may be used to determine the approximate mean and standard deviation of the series. (A best fit line is a straight line drawn through the centre of the points, so that the sum of the distances between the line and each point is at a minimum).

It will be seen that there are two vertical scales on probability paper, and that one is a mirror image of the other. The scale on the right shows the probability that values on the precipitation scale will be *below* that percentage. The scale on the left shows the probability that values on the precipitation scale will be *above* that percentage. For example, the 80% line on the right hand scale corresponds to the 20% line on the left hand scale. The value on the horizontal scale corresponding to the point where this line intersects the line of 'best fit' is about 96 cm. And so it is possible to predict that about 80 years in 100 will have precipitation totals of *less* than 96 cm, and thus 20 years in 100 will have precipitation totals of *more* than 96 cm.

It will be remembered that the arithmetic mean bisects the normal curve, and that 50% of the values fall above it, and 50% below. We should therefore expect that the horizontal line indicating 50% on the probability paper would approximately bisect our line of best fit. If we read where that intersection takes place on the horizontal scale, this should be the mean of the distribution. In fact the value of the mean taken from the graph is about 85, a close approximation to the calculated value of $\bar{x} = 84$.

It is also possible to determine the standard deviation from the graph. We know that 34% of values will be within one standard deviation above the mean, and 34% within one standard deviation below the mean. So if we inspect the 'best fit' line 34% below the mean (i.e. 34% below 50%) we

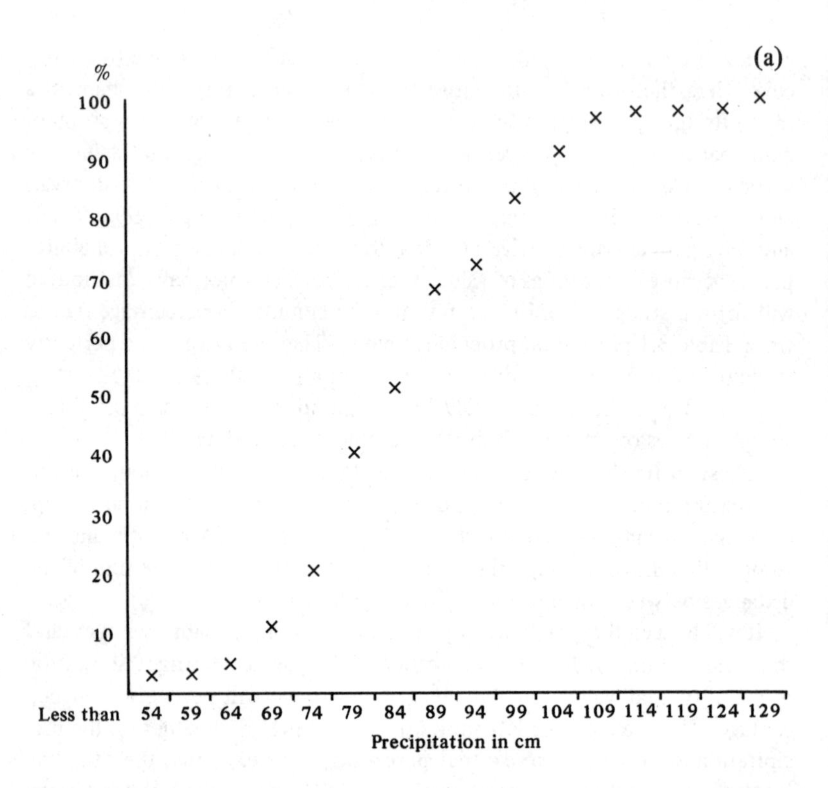

can read off one standard deviation on the horizontal scale. Reference to Fig. 3.2b shows that 34% below the mean on the 'best fit' line gives a value of about 70 on the precipitation scale. We already noted the mean as 85, so one standard deviation below the mean is $85 - 70 = 15$. Again this is a close approximation to our calculated value of $\sigma = 14{\cdot}5$ (grouped data).

Similarly it will be found that 34% above the mean yields a value of about 100 on the precipitation scale. Which again gives us a standard deviation of about 15 $(100 - 85 = 15)$.

Finally, probability paper is useful to estimate whether a given distribution approximates to normal. This is important as the characteristics of the standard deviation, which we have assumed, are valid only if the population distribution is approximately normal. To calculate how close a given distribution is to normal is not easy, but it may be estimated very quickly if the cumulative percentage frequency is plotted on probability paper.

In the above example, grouped frequencies have to be used for convenience, but of course all that has been said remains true if individual values

Fig. 3.2 *Cumulative percentage frequency graphs drawn from values given in Table 3.1. (a) shows the percentage cumulative frequency curve on ordinary graph paper; (b) shows the same values plotted on probability paper*

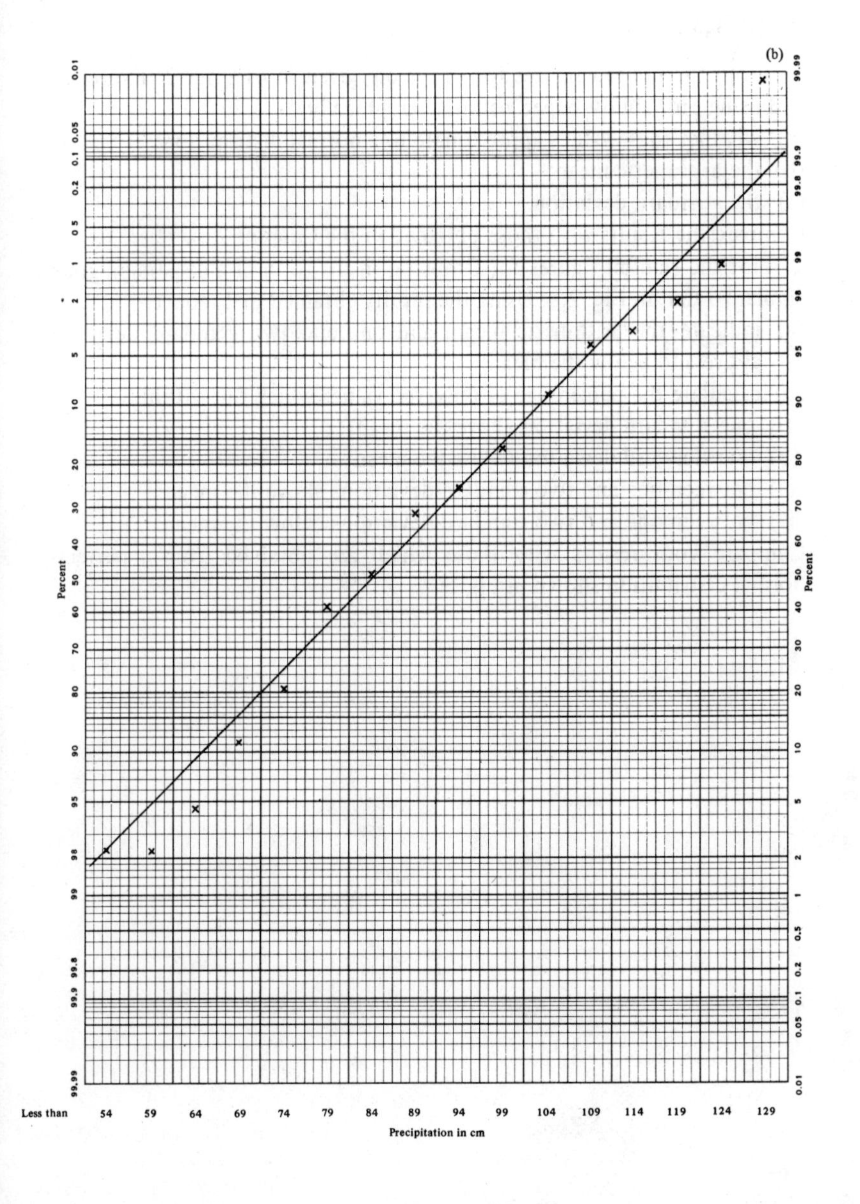

are used, and indeed individual values provide a more accurate result. The only difference in the calculation would be that actual values would take the place of groups, each value having a frequency of one.

Exercises

1. From the rainfall figures for Cropston, given in Chapter 2, Exercise 5, and using probability paper, estimate graphically:
(a) whether these values approximate to a normal distribution;
(b) the arithmetic mean;
(c) the standard deviation.
Compare the results with those obtained in the Chapter 2 exercise.
2. Plot the rainfall figures for Cropston in the form of a dispersion graph, similar to that in Fig. 2.8. Using the results obtained in 1. above:
(a) indicate the arithmetic mean, and one and two standard deviations above and below the mean;
(b) calculate the percentage of observed values which fall within:
(i) one standard deviation above the mean;
(ii) one standard deviation below the mean;
(iii) between one and two standard deviations above the mean;
(iv) between one and two standard deviations below the mean.
Compare your results with those which are obtained from a perfectly normal distribution shown in Fig. 2.7.

Chapter 4

Inferential tests

Classical and distribution-free statistics

In order to use the standard deviation predictively it is necessary that the background population should be distributed normally. Fig. 2.8 demonstrates this quite clearly. A little thought will show that a very much larger proportion of values clustered on one side of the mean (i.e. if the curve is not roughly symmetrical), would render prediction invalid. **All classical statistics, sometimes called parametric statistics, are based on the standard deviation and the properties of the normal curve, and require measurement on an interval scale.** An example is the product moment correlation coefficient used in Chapter 7.

But very often the kind of data which the geographer wishes to use may either not be normally distributed, or there is no means of knowing whether it is normally distributed or not. This is especially critical when only a small number of values is involved. Fortunately in recent years techniques have been evolved which do not have the requirement that the data should be normally distributed. These are known as **distribution-free statistics,** or sometimes **non-parametric statistics.** The new techniques are also useful, apart from being free from the restrictive assumption of normal distribution, because they:

1. **are effective with small samples;**
2. **may be used when only nominal or ordinal measurement is possible;**
3. **are generally much simpler and quicker to calculate** (important for results when doing fieldwork);
4. **can, in practice, be as powerful as their classical equivalent, provided that the sample size is increased slightly.** (Power efficiency is explained below in relation to the Mann Whitney U-test.)

The Mann Whitney U-test

This test is used to establish whether the difference in the means of two independent samples is statistically significant, that is, whether the difference is such that the two samples may be assumed to come from different populations. The test can be employed when measurement is on an ordinal scale, or when interval measurements have been placed in rank order. No assumption need be made concerning the distribution of the data, and it is effective with small, medium, and large samples.

Suppose that during the course of field work in a large urban area, it was decided, as part of the project, to try to establish whether prices for similar common food items were significantly different when purchased in the small shops of neighbourhood service areas, or in the large chain stores of the main shopping and business area (excluding the market). The method used was to make out one list of items and then record the price asked for each item in each large store, and at the appropriate shops in each local suburban shopping area. Table 4.1 gives the raw data.

Table 4.1 *Price in pence of a list of common food items sold in large stores (A), and in suburban neighbourhood service centres (B)*

A	174	179	169	170	173	
B	182	176	175	183	172	180

Food prices vary considerably seasonally, and sometimes between the beginning and end of the week. There are also considerable variations, especially in the smaller shops, depending upon the judgment of the individual shopkeeper. Table 4.1 seems to reveal evidence that the neighbourhood shops are slightly more expensive, with a mean cost of 178p, compared with a mean cost of 173p in the large stores, although goods from one shop were actually cheaper than those at three of the larger stores, and there is only a 3% difference between the means of the two samples. In such a case the Mann Whitney U-test can be used to establish whether the difference in cost is statistically significant. First it is necessary to formulate our hypotheses.

The null hypothesis, H_0, is *that there is no significant difference in terms of cost between the food items bought in shops and those purchased in large stores, and that the observed differences are due to chance variations.*

The alternative hypothesis, H_1, is *that the variations are too great to be the result of chance and the prices of the food items in the two samples are significantly different.*

We now have to decide upon our rejection level. If we choose the ·05 level of significance then there may be a 5% chance that we shall reject H_0 wrongly. That is, in taking our sample we have observed one of the rare occasions (a 1 in 20 probability) when the variations are due, perhaps, to a series of chance decisions by a number of individual shopkeepers, and our decision has been, wrongly, to accept H_1. However, to be able to reject H_0 with 95% confidence is generally acceptable in a problem of this nature (it would not be, for example, if we were testing a new drug), and so we set our rejection level at ·05, i.e. $\alpha = \cdot 05$.

We are now ready to apply the test.

The data are first re-arranged in a rank order sequence, with the group identity of each value retained. The samples are of different size, but this does not affect the validity of the test, the smaller sample being normally designated n_1. Thus in this case $A = n_1 = 5$, and $B = n_2 = 6$.

A	A	B	A	A	B	B	A	B	B	B
169	170	172	173	174	175	176	179	180	182	183

The statistic which we now have to find is termed U. It is obtained by inspecting both samples placed in a single rank order, and counting the number of times a B value *precedes* an A value, working from left to right. *All* the B values must be counted to the left of each A value. The initial two A values, 169 and 170, have no B to the left and so score 0. The first A value to have one B value preceding it is 173, thus A 173 scores 1. The next A value (A 174) has the same B value preceding it and therefore it also scores 1. The final A value to be preceded by a B value is A 179. This has B 176, B 175, and B 172 to the left, and so scores 3. U is simply the sum of these scores. That is $U = 0 + 0 + 1 + 1 + 3 = 5$.

Reference to Appendix 2 shows that with $n_1 = 5$, and $n_2 = 6$, the probability associated with a U value of 5 is ·041. Therefore the probability of the values occurring in samples A and B solely through random variation is ·041, or a situation that is likely to happen through chance only about 4 times in 100. And so we are able to reject H_0 at the ·05 level of significance, and accept H_1. To put it another way, we are able to say with 95% confidence, that the average price of the sample of food items sold in the large stores was significantly (in a statistical sense) lower than the same items sampled from the neighbourhood service centres.

The question now arises as to what would happen if, instead of counting the number of times a B value preceded an A value, we were to count the number of times an A value preceded a B value. Proceeding in exactly the same way that we did above, but counting the number of As preceding each B, the result would be as follows:

$$U = 2 + 4 + 4 + 5 + 5 + 5 = 25$$

On every occasion this test is used there will be two values obtainable for U. The lower value is known as U, but the higher value called U_1. However, if the higher value, U_1, is calculated it will be seen that it does not appear in the significance tables. If this happens U_1 may easily be transformed to U by the formula:

$$U = n_1 n_2 - U_1$$

Formula 4.1

In the above example $n_1 = 5$, and $n_2 = 6$. Therefore:

$$U = (5 \times 6) - 25$$
$$= 5$$

which is the value we have just found for U above.

The above example was given to demonstrate how the test works, and it is of course a perfectly satisfactory way of finding U when the samples are small. As the sample size increases it becomes more tedious to count individual As and Bs, and the possibility of error increases too. Fortunately, the value of U may also be calculated by using the following formula:

$$U = n_1 n_2 + \frac{n_1(n_1 + 1)}{2} - R_1$$

Formula 4.2a

or:

$$U = n_1 n_2 + \frac{n_2(n_2 + 1)}{2} - R_2$$

Formula 4.2b

where n_1 is the number of values in the smaller sample, n_2 is the number of values in the larger sample, and R_1, R_2 are the sums of a single sequence of *rank* values in n_1 and n_2 respectively (see worked example, Table 4.2).

A good example of the value of using a statistical test is the comparison of wheat *yield* between northern and southern England. Table 4.2 shows average wheat yields in kg/hectares for the years 1966–70 for selected counties. It will be seen that the mean yield for the northern counties is 678 kg/ha (kilogram per hectare) while the mean for the southern counties is 633 kg/ha. On the whole, conditions for growing wheat in the north differ little from the south. For this reason it might be anticipated that yield would be roughly the same, although it might vary a little from year to year. What slight variation there is climatically would seem marginally to favour the south. In fact, it will be seen from Table 4.2 that there is only a 6% difference in the mean yield between the two areas, but that, perhaps surprisingly, the lower yield was in the south.

It is tempting to conclude that a difference of this magnitude, and in this direction, might well be the result of chance variations, occurring during the years in which the yield is compared. Before we can attempt to draw conclusions based on the evidence, it is necessary to calculate the probability that differences in yield are the result of random variation. It is exactly this kind of information that a statistical test can provide.

Table 4.2 *Wheat yield in kg/ha in selected counties in southern and northern England for the years 1966–70*

(1) Northern counties $n = 9$		Ranks of 1	(2) Southern counties $n = 13$		Ranks of 2
Northumberland	755	22	Hampshire	658	15
Cumberland	707	20	I. of W.	654	13
Westmorland	722	21	Berkshire	638	8
Lancashire	646	10	Oxfordshire	631	6
Yorks:			Buckinghamshire	605	3
E.Riding	697	19	Sussex (E)	615	4
N. Riding	672	17	Sussex (W)	652	11
W. Riding	629	5	Kent	640	9
Durham	681	18	Surrey	545	1
Lincs: Lindsey	590	2	Wiltshire	660	16
			Gloucestershire	653	12
			Somerset	655	14
			Dorset	634	7
		$\Sigma R_1 = 134$			$\Sigma R_2 = 119$
$\bar{x}_{n_1} = 678$			$\bar{x}_{n_2} = 633$		

First we must formulate our hypotheses. H_0, the null hypothesis, states *that there is no significant difference between the two sample means, and that the observed differences between the sample values are the result of chance variations.* H_1, the alternative hypothesis, states *that the two sample means differ significantly, because the differences between the values in each sample are unlikely to be the result of chance.*

The **rejection level** may reasonably be put in this case at $\alpha = \cdot 05$, that is, we shall be content to reject H_0 provided there is a 95% probability that the variations cannot be ascribed to chance.

The Mann Whitney test is well suited to establish whether the differences in the two samples are great enough to be statistically significant. Once again the data have to be ranked overall (i.e. the rank numbers are in a consecutive sequence through both samples), but this time the samples themselves are kept separate. It is usual in the U-test to give the rank of 1 to the data with the lowest value. (If the highest value is given the rank of 1 then U and U_1 are reversed.) The rank values (R) are then summed. Our data for the test, from Table 4.2, are as follows:

$n_1 = 9, \ R_1 = 134,$ $n_2 = 13, \ R_2 = 119.$

We are now able to apply Formula 4.2a

$$
\begin{aligned}
U &= n_1 n_2 + \frac{n_1(n_1 + 1)}{2} - R_1 \\
&= 9 \times 13 + \frac{9(9+1)}{2} - 134 \\
&= 117 + 45 - 134 \\
&= 28
\end{aligned}
$$

If we use Formula 4.2b the result will be:

$$
\begin{aligned}
U &= n_1 n_2 + \frac{n_2(n_2 + 1)}{2} - R_2 \\
&= 9 \times 13 + \frac{13(13+1)}{2} - 119 \\
&= 117 + 91 - 119 \\
&= 89
\end{aligned}
$$

89 is the larger value of U we have referred to as U_1. This can be proved by applying Formula 4.1:

$$U = n_1 n_2 - U_1 = 117 - 89 = 28.$$

It is useful to apply this formula as the answer is a check that there has been no error in the arithmetic.

It is now necessary to test the U value obtained to see whether it is significant. Reference to Appendix 2 shows that with $n_1 = 9$ and $n_2 = 13$, the highest value of U that is significant at the ·05 level is 28. We can therefore reject H_0, accept H_1, and conclude that, despite the fact there is no great difference between their respective arithmetic means, the variation in the values contained in each of the two samples is so great that we can say with 95% confidence they are drawn from different populations. The difference in wheat yield between the northern and southern counties is thus statistically significant at the ·05 level. (Note that with the U-test a calculated value of U equal to *or lower than* that given in the Appendix is required for the result to be significant at the stated level.)

The significance of this result may be a little surprising, and it makes a very important point. So often in the past geographers have concluded that untested relationships or differences were 'significant' simply by a visual comparison of mean values, tables of figures, or distributions plotted on maps, and have thus lacked rigour in their work. This stage of geography is now over. What must also have happened in the past, as exemplified by the results above, is that *less striking differences were often missed that were, in fact, statistically significant.*

So far the U-test has been used in cases where each of the two samples contained less than twenty values ($n_1 \leqslant n_2 \leqslant 20$). But if either sample *exceeds* twenty it is possible to calculate a z-score and determine the significance of U from the normal distribution table in Appendix 1. For example, during the course of field work in South Devon, a problem arose concerning 'head' material exposed on two sites along the top of sections of sea cliff near Budleigh Salterton. The problem was whether the material on each of the two sites had a common origin. This appeared possible because of the configuration of the land, and because in samples from both sites the 'head' contained large quantities of quartz pebbles. Further inland the Bunter Pebble beds outcrop over a considerable area, and this seemed a possible and likely place for the pebbles to come from.

To test this hypothesis, random samples of quartz pebbles were taken from both sites, and a coefficient of roundness calculated for each. If pebbles are formed by a similar process from the same material, then the extent to which they are rounded will depend on their erosional history. (For example, if two comparable pieces are detached from the same parent rock by mechanical weathering, and fall into a river, and one is rapidly covered by fluvial sediments, while the other spends a long period being transported downstream as part of the river's bedload, it is evident that the latter, because it is being rolled along the bed of the stream, will become much more rounded than the former, which has been protected from further erosion.)

The coefficient of roundness (or sphericity), R, may be calculated from the following simple formula:

$$R = \frac{2r}{L} \times 100$$

where L is the length of the long axis of the major plane of the pebble, and r is the radius of the arc forming the sharpest curve in the major plane (see Fig. 4.1).

Fig. 4.1 *The long axis and angle of sharpest curvature in the major plane, of a typical pebble. In this case, the circle of which this arc of curvature of the sharpest angle is a part has a radius of 12mm, and the length of the long axis is 54mm. The coefficient of roundness, therefore, is approximately* $R = \frac{2 \times 12}{54} \times 100 = 44$. *The technique is crude, but has been found very effective in practice. The formula yields convenient values up to 100, when the stone is spherical. (In cases where this range is deemed not to be sufficiently discriminating, 1000 is normally substituted for 100, giving a range of values up to 1000)*

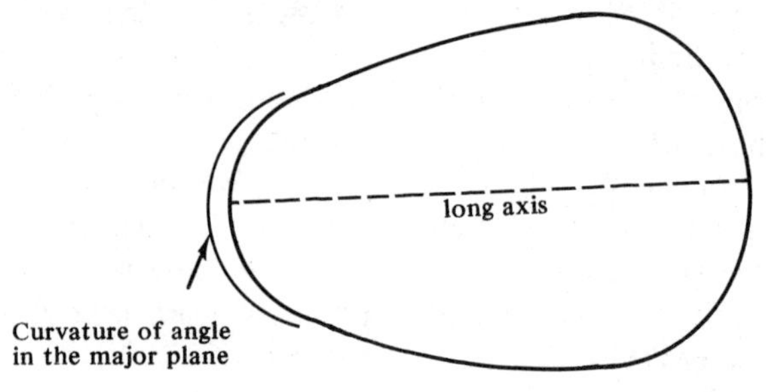

The method of transport of the pebbles to both sites was similar, i.e., by solifluction within the head material. If the samples had roundness coefficients that were significantly different then they must have either a) originated from different locations on the Pebble Beds, or b) come from the same location, but at a different period of time (when a different horizon of the Pebble Beds was exposed to periglacial activity).

Table 4.3a *Coefficients of roundness of pebbles from two sites on the cliffs of the South Devon coast*

	57	47	33	10	20	19			
Sample A	8	9	11	14	28	22	$\bar{x} = 22$		
	14	14	15	19	29	30			
	34	12	58	39	25	14	18	52	
Sample B	7	27	56	42	14	21	46	48	$\bar{x} = 33$
	31	13	45	53	24	16	48	49	

The coefficients of pebble roundness actually measured are given in Table 4.3a. From inspection it will be seen that Sample B has a number of pebbles with higher coefficient values than many in Sample A. This is borne out by the respective mean values of $\bar{x} = 22$, and $\bar{x} = 33$. Yet Sample B contains not only the roundest pebble (58), but also the most angular (7). A statistical test is required to determine whether the difference between the two samples is significant, and thus whether we may regard the roundness values in the two samples as being so different that they are unlikely to be drawn from the same parent population.

First, we must formulate our hypotheses. H_0 states *that, in terms of stone roundness, the two samples are drawn from the same population, and any differences are the result of random variation.* The alternative hypothesis, H_1, is *that the differences in stone roundness values are so great that they are unlikely to be the result of chance, and the two samples must therefore have been drawn from different populations.*

Table 4.3b *Ranks of sample values given in Table 4.3a*

Sample A	Rank 1	Sample B	Rank 2
57	41	58	42
47	34	56	40
33	28	53	39
30	26	52	38
29	25	49	37
28	24	48	35·5
22	20	48	35·5
20	18	46	33
19	16·5	45	32
19	16·5	42	31
15	13	39	30
14	10	34	29
14	10	31	27
14	10	27	23
11	5	25	22
10	4	24	21
9	3	21	19
8	2	18	15
	$\Sigma R_1 = 306$	16	14
		14	10
		14	10
		13	7
		12	6
		7	1
			$\Sigma R_2 = 597$

Our rejection level is decided upon as $\alpha = \cdot 05$.

To calculate U it is again necessary, as in the previous example, to place all the values in a single rank order, and to maintain the identity of each sample. This is given in Table 4.3b.

Action to be taken with ties

It will be seen that certain pebbles in each sample have the same roundness value, e.g. there are two 19s in Sample A; and one value, 14, occurs three times in Sample A, and twice in Sample B. These are called 'tied' values. (Theoretically, if measurements are made sufficiently accurately no pebble will have *precisely* the same roundness coefficient as another. In practice measuring techniques cannot give this degree of accuracy, or the effort involved may be considered too laborious and time-consuming.) When there are two or more variates which, in practice, are allotted the same value, they are given the *average* of ranks they might otherwise have occupied. For example, in Sample A two pebbles have coefficients of 19. Now the coefficient 18 in Sample B has a rank of 15, so a coefficient of 19 would normally be given the rank of 16. But as we have *two* 19s they are allotted the *average* of ranks 16 *and* 17, i.e., 16·5. Similarly, five pebbles have a coefficient of 14, three of Sample A and two in Sample B. All five are allotted the average of ranks 8, 9, 10, 11, and 12, i.e. 10.

It is now possible to find the value of U using Formula 4.2a. Let Sample A be n_1, and Sample B n_2.

$$U = n_1 n_2 + \frac{n_1(n_1 + 1)}{2} - R_1$$

$$= (18 \times 24) + \frac{18(18 + 1)}{2} - 306$$

$$= 432 + 171 - 306$$

$$= 297$$

Alternatively we could use Formula 4.2b, in which case:

$$U = n_1 n_2 + \frac{n_2(n_2 + 1)}{2} - R_2$$

$$= (18 \times 24) + \frac{24(24 + 1)}{2} - 597$$

$$= 432 + 300 - 597$$

$$= 135$$

The values for U will differ depending on whether we use Formula 4.2a or 4.2b. Formula 4.1 is a useful check on arithmetic, although it is not essential.

$$\begin{aligned} U &= n_1 n_2 - U_1 \\ &= (18 \times 24) - 135 \\ &= 432 - 135 \\ &= 297 \end{aligned}$$

To find the level of significance of U with $n > 20$ we are able to calculate a z-score using Formula 4.3.

$$z = \frac{U - \frac{n_1 n_2}{2}}{\sqrt{\frac{n_1 n_2 (n_1 + n_2 + 1)}{12}}}$$

Formula 4.3

Substituting the known values given above in Formula 4.3,

$$z = \frac{297 - \frac{(18 \times 24)}{2}}{\sqrt{\frac{18 \times 24 (18 + 24 + 1)}{12}}}$$

$$= \frac{297 - 216}{\sqrt{\frac{432 \times 43}{12}}} = \frac{81}{39{\cdot}35} = 2{\cdot}1$$

It does not matter whether U = 297 or U = 135 is inserted in Formula 4.3. Numerically the result will be the same. The difference is that if the lower value of U is used the sign of the z-score will be negative. This is to be expected, because z relates to the symmetrical normal curve, and a positive score simply indicates the area above the mean, while a negative score indicates the area below the mean.

Reference to Appendix 1 shows that a z-score value of 2·1 has an associated probability of about ·036. (For U = 135 with a z-score of −2·1 the probability is ·018, and for U = 297 with a z-score of +2·1 the probability is also ·018. The *total* probability therefore is ·018 + ·018 = ·036).

Our test has thus shown that the probability that the two sets of sample values are part of the same population (i.e., that differences are due to chance variations), is only ·036, or 3·6%. The level of rejection chosen was $\alpha = \cdot05$. (In other words we were prepared to reject H_0 provided there was less than a 5% chance that it might be true.) We are thus able to reject H_0 with more than 95% confidence and conclude that, in terms of pebble roundness, our two samples are drawn from different populations.

It follows that if the differences in roundness coefficients are unlikely to be due to chance then the pebbles in the two samples must have had a different erosional history. As the soliflual head material in which they were embedded appeared to have provided identical means of transport to the present sites of both samples, then, either the pebbles originated from a different location, or they came from the same place but at a different time and from a different horizon in the Bunter Sandstone.

Tied ranks do not greatly affect the outcome of the U test, unless there are very many of them, or unless the same value occurs in a long 'run'. Fortunately, the existence of ties tends to *increase* the rigour of the test, that is, their presence *tends to reduce the value of z,* and thus reduces the probability that H_0 will be wrongly rejected. If the required rejection level is reached, therefore, the result of the test may be accepted irrespective of the number of ties involved.

Power efficiency

Tests using interval measurement and based on the assumption that the data is normally distributed (i.e. classical statistics based on the properties of the normal curve) are generally rather more 'powerful' than distribution-free tests. By **power** we mean the ability of the test correctly to reject H_0. Student's *t*-test is the most powerful classical method for testing the significance of the difference between the means of two independent samples. (The *t*-test is not considered in this book, because of its lengthy calculations and the restrictive assumptions regarding its use. The far simpler U-test is an excellent alternative.) The Mann Whitney U-test performs exactly the same function, but uses data transformed into ordinal (ranked) values, thus 'throwing away' some of the information if the original data is on an interval scale. The U-test is therefore less powerful (less effective) than the *t*-test, when all the assumptions necessary for the *t*-test have been met. If the *t*-test is considered as the standard having 100% power, then the U-test is 95% as effective. In practice this slight disadvantage can easily be overcome by increasing the size of the samples to make up for the information lost in using data in an ordinal form. For example, the *t*-test and the U-test both have equal power when the number of values in the samples are 95 and 100 respectively.

There are many occasions when it is not possible to use a classical test, because data is not on an interval scale, or a normal distribution of the population cannot be assumed. In these circumstances (which often occur in geographical work), and when no other test is available for comparison, power efficiency is not meaningful.

The chi-squared test for *k* independent samples

So far we have used χ^2 to test a single distribution (p. 6), and the U-test to determine whether a significant difference exists between two independent samples. Cases often occur, however, when we wish to establish the significance of differences between more than two groups of observations. For example, in many areas of Highland Britain enclosed fields, sometimes of considerable antiquity, are presently being abandoned in what appear to be marginal areas. Sometimes the rock type and the characteristics of the soil which develops upon it may be a contributory factor. Often altitude and aspect, in otherwise relatively homogeneous areas, seem to be of considerable importance. Sometimes the pattern appears to be quite random, resulting from the choice of many individuals between greater economic gain in industry, and the love of a traditional life on the land.

Let us suppose we have a relatively homogeneous area with respect to the economic advantages of soil type and markets for farm products, and examine the pattern of abandoned fields, choosing aspect as the attribute in question. First we must define our terms.

1. Flat land does not have 'aspect', so we define 'slope' as land having an angle of 5° or more.
2. Aspect is a continuum through 360°, but probably the most important directions from the point of insolation and other climatic differences are North, South, East, and West. 'North' aspect is defined as a field on a slope facing between north east and north west and so on.
3. A field where there is any doubt concerning either (1) or (2) (e.g. if part of a field has less than a 5° angle of slope) should be discounted.
4. Altitude is also measured on a continuous scale, and some arbitrary division is necessary to decide the height interval into which the frequencies of abandoned fields will be put (a scattergram in which the fields are plotted against altitude may be of help in deciding the interval). In this case a suitable interval of 50 m, from 100 m to 300 m, is chosen.
5. Holdings are continually being abandoned, in the sense that their owners leave them, either by dying or by moving elsewhere. But if they are favour-

ably situated the land is sold and kept cultivated. A field is only considered abandoned when no buyer is forthcoming and the land allowed to revert to the wild vegetation of the area.

The number and aspect of abandoned fields which result from our researches are as follows:

Table 4.4a *Aspect and frequency of abandoned fields in sample area*

Altitude in metres	k_1 North	k_2 South	k_3 East	k_4 West	Σr
r_1 100 – 150	13	0	14	1	28
r_2 151 – 200	21	2	17	9	49
r_3 201 – 250	20	9	21	10	60
r_4 251 – 300	30	13	29	16	88
Σk	84	24	81	36	$N = 225$

Data set out like this are said to form a **contingency table**, in which the columns are conventionally called $k_1, k_2 \ldots k_n$, and the rows $r_1, r_2 \ldots r_n$.

Inspection of Table 4.4a shows an overall increase with altitude of the abandoned fields, but considerable differences in number between one *aspect* and another. As one would expect, the most favourable aspect for agriculture lies between south-east and north-west, and thus the number of fields contained in k_2 and k_4 total only 60, whereas of those facing in a northerly or easterly direction, 165 have been abandoned. From this evidence there seems little doubt that aspect is a highly important factor in the decision to allow fields to revert, probably because yield is affected by the rather less favourable climatic conditions prevailing on north and east facing slopes, especially when combined with increasing altitude. But before committing ourselves to a definite statement, it is necessary to know with what degree of confidence we can make the assertion. For this it is necessary first to formulate our hypotheses, and to decide upon the rejection level of H_0.

H_0 is *that fields are abandoned irrespective of altitude and aspect, and observed frequencies are the result of individual whims and thus represent chance variations.*

H_1 is *that altitude and aspect significantly affect individual decisions to abandon the more marginally productive fields.*

Rejection level is decided at $\alpha = {\cdot}05$.

The kind of test chosen depends upon the nature of the available information. In our case the data consists of *frequencies* (of fields) 'measured' on a nominal scale (the four aspect categories), against an ordered sequence (the intervals of altitude). As measurement is at a nominal level, and in the form of frequencies, a χ^2 test seems indicated. The number of fields abandoned within each altitude interval for every aspect could obviously form the observed frequency for every fraction. What is required also is to calculate an expected frequency appropriate to each, that is the frequency which might be expected if the association were random. To see how this may be done, the information is repeated in Table 4.4b.

Table 4.4b *Expected frequencies for the data set out in Table 4.4a*

		k_1	k_2	k_3	k_4	
r_1	O	13	0	14	1	*28*
	E	*10·45*	*2·99*	*10·08*	*4·48*	
r_2	O	21	2	17	9	49
	E	*18·29*	*5·23*	*17·64*	*7·84*	
r_3	O	20	9	21	10	60
	E	*22·40*	*6·40*	*21·60*	*9·60*	
r_4	O	30	13	29	16	88
	E	*32·85*	*9·39*	*31·68*	*14·08*	
		84	24	81	36	$N = 225$

To determine the expected frequency for any cell in the contingency table (i.e. the frequency expected under H_0), all that is required is to multiply the sum of the row (r) in which the cell occurs by the sum of the column (k) in which it occurs and divide by N, the total observed frequencies. For example, we wish to obtain the expected frequency appropriate to the observed frequency of 13 in the top left hand cell. The sum of r_1 is 28. The sum of k_1 is 84. The total of observed frequencies is 225. Therefore the expected frequency (E) we require is:

$$E = \frac{\Sigma r_1 \times \Sigma k_1}{N}$$

$$= \frac{28 \times 84}{225}$$

$$= 10{\cdot}45$$

Table 4.4b shows the expected frequency (printed in italic figures) calculated for each cell.

To use a χ^2 test on a contingency table of this kind, the requirements are the same as for the χ^2 test used previously (p. 12). Briefly, the information must be in the form of frequencies; total *observed* frequencies must be at least 20; no cell may have an *expected* frequency less than one, and at least 80% of cells must have an *expected* frequency of five or more. In this case a χ^2 test is permissible.

The formula for finding χ^2 when used with a contingency table is:

$$\chi^2 = \Sigma^r \Sigma^k \frac{(O-E)^2}{E}$$

Formula 4.4

where O is the observed frequency in each cell, E is the expected frequency in each cell, and $\Sigma^r \Sigma^k$ means sum over all rows and columns, that is, sum the values of $\frac{(O-E)^2}{E}$ for each cell.

Using Formula 4.4 to obtain the χ^2 value for the data in Table 4.4b:

$$\begin{aligned}\chi^2 &= \Sigma^r \Sigma^k \frac{(O-E)^2}{E} \\ &= \frac{(13-10{\cdot}45)^2}{10{\cdot}45} + \frac{(0-2{\cdot}99)^2}{2{\cdot}99} + \frac{(14-10{\cdot}08)^2}{10{\cdot}08} + \frac{(1-4{\cdot}48)^2}{4{\cdot}48} \\ &+ \frac{(21-18{\cdot}29)^2}{18{\cdot}29} + \frac{(2-5{\cdot}23)^2}{5{\cdot}23} + \frac{(17-17{\cdot}64)^2}{17{\cdot}64} + \frac{(9-7{\cdot}84)^2}{7{\cdot}84} \\ &+ \frac{(20-22{\cdot}40)^2}{22{\cdot}40} + \frac{(9-6{\cdot}40)^2}{6{\cdot}40} + \frac{(21-21{\cdot}60)^2}{21{\cdot}60} + \frac{(10-9{\cdot}60)^2}{9{\cdot}60} \\ &+ \frac{(30-32{\cdot}85)^2}{32{\cdot}85} + \frac{(13-9{\cdot}39)^2}{9{\cdot}39} + \frac{(29-31{\cdot}68)^2}{31{\cdot}68} + \frac{(16-14{\cdot}08)^2}{14{\cdot}08} \\ &= 13{\cdot}90.\end{aligned}$$

When a chi-squared test was used in Chapter 2, the calculated value of χ^2 was tested by using a graph which showed the number of times out of 100 the observed distribution might be expected to occur through chance. This simplification was used so that a simple χ^2 test could be introduced without lengthy explanations concerning the properties of the normal curve and levels of significance. But *any* calculated value of χ^2 (with the appropriate degrees of freedom) may be tested for significance simply by using the table of critical values of chi-squared in Appendix 3. First it is necessary to establish the degrees of freedom involved. In the case of a contingency table they are found by multiplying the number of rows less one, by the number of columns less one. In this case:

$$df = (r-1)(k-1) = (4-1)(4-1) = 9.$$

We are now able to test our result of χ^2 = 13·90 (with df = 9) from Appendix 3. Provided that 13·90 is greater than the critical value of chi-squared with 9 degrees of freedom at the ·05 level of significance, we will be able to reject H_0. In fact, with $df = 9$, and α = ·05, it will be seen from Appendix 3 that the critical value of chi-squared is 16·92. Statistically there is more than a 5% chance that the distribution of abandoned fields we have observed is the result of random variations.

When we look at Table 4.4a in conjunction with the alternative hypothesis H_1, this seems a surprising result. Before the adoption of measures of statistical rigour most geographers would have confidently asserted that, on the information available, aspect affected individual decisions to abandon fields. They might well have been correct, but with a known 5% possibility of error, the assertion is unlikely to be made today. If aspect is to be claimed as a causal factor, more information is required until H_0 can be rejected at the required level of significance.

This method of using χ^2 may be applied to a contingency table containing any number of rows and columns, though the calculation becomes rather laborious, unless a calculator or computer is used, if the number of cells is very great.

It must be remembered that, in the above example, we investigated the influence of aspect through the whole circle of 360°, and we chose the four cardinal points for our test. The result was not significant, but it is possible that it might have been so had different directions been selected, or the differences between more or fewer aspects examined. Suppose we extract from Table 4.4a the information we have regarding the number of north-facing and south-facing abandoned fields, set it out as Table 4.4c,

Table 4.4c *Expected frequencies derived from Table 4.4a, but restricted to north and south facing fields only*

Altitude in metres			k_1 North	k_2 South	Σr
100 – 150	r_1	O	13	0	13
		E	*10·11*	*2·89*	
151 – 200	r_2	O	21	2	23
		E	*17·89*	*5·11*	
201 – 250	r_3	O	20	9	29
		E	*22·56*	*6·44*	
251 – 300	r_4	O	30	13	43
		E	*33·44*	*9·56*	
		Σk	84	24	$N = 108$

and submit it to a χ^2 test in exactly the same way as before. Our hypotheses and level of rejection remain the same, although this time aspect refers only to north and south.

The expected frequency for each cell is calculated as before, being the sum of the row multiplied by the sum of the column, divided by the total observed frequencies. Thus the expected frequency for the cell containing 21 in r_2 and k_1 is:

$$\frac{\Sigma r_2 \times \Sigma k_1}{N} = \frac{23 \times 84}{108} = 17{\cdot}89.$$

The value of χ^2 is now obtained using Formula 4.4.

$$\chi^2 = \Sigma^r \Sigma^k \frac{(O - E)^2}{E}$$

$$= \frac{(13 - 10{\cdot}11)^2}{10{\cdot}11} + \frac{(0 - 2{\cdot}89)^2}{2{\cdot}89} + \frac{(21 - 17{\cdot}89)^2}{17{\cdot}89} + \frac{(2 - 5{\cdot}11)^2}{5{\cdot}11}$$

$$+ \frac{(20 - 22{\cdot}56)^2}{22{\cdot}56} + \frac{(9 - 6{\cdot}44)^2}{6{\cdot}44} + \frac{(30 - 33{\cdot}44)^2}{33{\cdot}44} + \frac{(13 - 9{\cdot}56)^2}{9{\cdot}56}$$

$$= 9{\cdot}05$$

Degrees of freedom, in the case of this contingency table, are $df = (r - 1)(k - 1) = (4 - 1)(2 - 1) = 3 \times 1 = 3$. Reference to Appendix 3 will show that with $df = 3$ the critical value of χ^2 is 7·82 at the ·05 level, and 11·34 at the ·01 level of significance. Our calculated value was $\chi^2 = 9{\cdot}05$. The rejection level was set at $\alpha = {\cdot}05$. We may therefore reject H_0, and accept the alternative hypothesis H_1.

It will be seen that when we test *part* of the data, and we have chosen the two aspects where the greatest difference exists, we obtain a result that is significant, and are able to state with 95% confidence that in this case the difference between facing north or south affects individual decisions to abandon fields.

It might perhaps be worth noting the large differences that can occur within a contingency table before a situation is reached when they are unlikely to be the result of chance. When all four cardinal aspects were considered, the different numbers of abandoned fields between north and east on the one hand, and south and west on the other (Table 4.4a), at first sight seemed sufficiently large to make very slight the probability that such a distribution was the result of chance variations. Yet the test showed this impression was false. Even greater relative differences existed between frequencies concerned with north or south facing slopes alone. It is true that the test showed these to be significant, but still only with 95%, rather than 99%, confidence.

A simple runs test

If a 'fair' coin is tossed in the air there is a 50% chance that it will fall heads uppermost, and a 50% chance that it will fall tails uppermost. If it is tossed in the same way a large number of times, then the number of 'heads' will equal the number of 'tails' almost exactly. But it will not be possible to predict the result on any specific occasion because at every toss the probability of a head remains 50%, although there will be a number of times when a sequence of several 'heads' or 'tails' will occur as the result of random variations in the process of tossing the coin. If as a result of 100 tries the first 50 fall 'heads' and the second 50 fall 'tails', we should rightly infer that the probability of this happening by chance was extremely remote, and conclude that something was influencing the fall of the coin. On the other hand, it is equally unlikely that the coin would invariably fall alternatively heads then tails 100 times.

Let us represent a 'head' by + and a 'tail' by –. Suppose we tossed a coin 25 times with this recorded result:

+ – + – – – + – + + – + –– + – + –– ++ – + – +

To determine whether this series is likely to have happened by chance, we must first count the number of 'runs', that is, the number of times a plus or minus is recorded on its own, or beside other plus signs or minus signs. In either case each consecutive series of pluses or minuses is termed a 'run'. The procedure is clear if we rewrite the sequence with the cumulative number of runs (r) marked below the symbols.

	+	–	+	–––	+	–	++	–	+	––	+	–	+	––	++	–	+	–	+
r	1	2	3	4	5	6	7	8	9	10	11	12	13	14	15	16	17	18	19

We have here about the same number of heads and tails, + = 12 and – = 13, arranged in a series having 19 runs ($r = 19$). Let the 12 pluses be termed n_1, and the 13 minuses n_2. It is possible to calculate the probability of obtaining, through chance variations, any number of runs for given sample sizes of n_1 and n_2. Appendix 4 shows that with $n_1 = 12$ and $n_2 = 13$, $r = 8$ *or less,* and $r = 19$ *or more* are significant at the ·05 level. In other words with samples of that size the probability of obtaining less than 9 runs or more than 18 runs through random variations is less than 5%.

We have so far been talking of tossing coins, but **this method can be applied to any data that can be reduced to 'Yes/No' terms.** Notice that it is not the *number* of + and – signs that matter, but *their order.* One of the techniques frequently used in geography is that of the transect. Here the thing that is important is **the order in which different phenomena occur.** The runs test is therefore suitable to assess whether or not the incidence of a particular object of study is distributed randomly along a line of transect, *provided that the data can be represented in an 'either/or' fashion.* In a urban area this might be whether a building was a dwelling house or not. In a rural area it might be whether there was a signification variation in land use along a line of transect by recording whether each field traversed was arable or not. That is, we could test whether the arable land was–more or less–evenly distributed, or whether the arable fields tended to cluster significantly in certain areas, indicating perhaps that soil, aspect, or some other feature made arable farming more profitable in those places.

The runs test may be applied to any data in which the *order* of occurrence is to be examined. Changes through time may be tested to determine whether cyclical fluctuations are significant. It is possible to attempt to distinguish small variations in some aspects of climate in this way. An example might be the annual rainfall figures recorded at Chatsworth from 1878 to 1924. We have already seen (p. 20) that the mean of all the recorded annual rainfall totals from 1878 to 1971 is 84 cm. If we now refer to the annual totals between 1878 and 1924 and allot a plus (+) for each year above 84 cm, and a minus (–) for each year below 84 cm, we have a situation in which a number of groups of years of high rainfall appear, together with groups of years of low rainfall (Table 4.5). For instance, the years 1878–1881 were well above the mean; 1895–1899 were below; 1909 to 1916 were all above the mean, except for one year; and 1917 to 1923 were all well below the mean. The problem was to discover whether this was a sufficiently strong relationship, or grouping, of years with high and low rainfall to infer that there might be some factor inducing small cyclical changes in the rainfall regime, or whether the observed groups might be expected through random variations in climate from year to year over a long period of time. For this a statistical test was necessary.

The null hypothesis, H_0, was *that each year's rainfall was independent of any other year, and that apparent 'runs' of years in which rainfall was high or low were the result of chance.*

H_1 was *that the number of runs was unlikely to be the result of chance, and therefore some factor was affecting rainfall for short periods of years.*

The rejection level could be $\alpha = {\cdot}05$.

Table 4.5 *Totals of annual precipitation in cm at Chatsworth 1878 to 1923 (4 years are missing)*

111	109	127	119	83	52	74	89
+	+	+	+	−	−	−	+
75	102	85	72	86			
−	+	+	−	+			
77	77	82	76	82	96	79	75
−	−	−	−	−	+	−	−
102	72	75	87	93	75	85	92
+	−	−	+	+	−	+	+
69	122	91	104	103	96		
−	+	+	+	+	+		
74	79	75	71	50	69	81	
−	−	−	−	−	−	−	

Let years with $>$ 84 cm precipitation (+) be n_1, and years with $<$ 84 cm precipitation (−) be n_2. Then $n_1 = 19$, $n_2 = 23$, and the number of runs observed (r) is 18.

The data are suitable for the runs test, but it will be observed that $n_2 = 23$, and Appendix 4 only deals in sample sizes up to $n = 20$. This is because if either sample exceeds 20 it is possible to establish the associated probability under H_0 by calculating a z-score. The formula for doing this looks formidable, but in fact is simple enough to use.

$$z = \frac{r - \left(\dfrac{2n_1n_2}{n_1 + n_2} + 1\right)}{\sqrt{\dfrac{2n_1n_2(2n_1n_2 - n_1 - n_2)}{(n_1 + n_2)^2\,(n_1 + n_2 - 1)}}}$$

Formula 4.5

where r is the number of runs observed, n_1 is the number of pluses, and n_2 the number of minuses.

From Table 4.5 we see that in our case $r = 18$, $n_1 = 19$, and $n_2 = 23$. Substituting these values in Formula 4.5 to obtain a z-score, we have:

$$z = \frac{18 - \left(\dfrac{2 \times 19 \times 23}{19 + 23} + 1\right)}{\sqrt{\dfrac{2 \times 19 \times 23(2 \times 19 \times 23 - 19 - 23)}{(19 + 23)^2\,(19 + 23 - 1)}}} = -1{\cdot}2$$

The z-score is negative because the number of runs observed is less than the expected mean number of runs with samples of this size. Appendix 1 shows that with $z = 1{\cdot}2$ the associated probability is ·230. In other words there is a 23% probability that H_0 is correct. So although it might be tempting to speculate about the weather pattern at Chatsworth, the probability that runs of years of high and low annual rainfall are due to chance variations is so great that any argument must have a base which is statistically unsound.

Paired data

The Wilcoxon rank test

So far we have dealt with data which come from *independent* samples. Sometimes we want to test for significance differences between paired values. In the example of difference in wheat yields between northern and southern counties in England (p. 41) the samples were independent, that is, they were not in any way related, so that one was uninfluenced by the other. However, suppose we had taken two samples *from each of the same counties, but in different years,* and then desired to find out whether better seed, better agricultural techniques, or some other factor had resulted in a significant difference in yield, we should be dealing with **paired values,** because the areas from which the sample was taken were the same on each occasion. Data concerning any phenomenon at the same place, but at a different time, must be regarded as paired. The changing angles of slope on beach profiles from season to season may be regarded as paired (provided the same profiles are measured for comparison each time), and the employment structure in a city may be regarded as paired if comparison is made between two census dates. An example of paired values, where time is not involved, might be the comparison of the slope angles of either side of a series of cross sections of a valley (in homogeneous material).

If data are paired, and therefore not independent, the values must be treated in a particular way. The Wilcoxon rank test is a distribution-free test that may be used with small samples ($n = 6$ to $n = 25$), or with large samples ($n > 25$), when 'matched' or paired data is involved.

Let us first take a small sample, and compare traffic flow, in terms of vehicles per hour, at ten check points on different radial roads leading out of a city (Table 4.6). The check is at the same place and time, but with an interval of five years between checks, during which time a by-pass round the city has been constructed. The problem is to find out whether, despite the by-pass, the traffic flow into the city has changed significantly.

H_0 is *that no change in the volume of traffic has taken place, and that the observed variations are the result of chance.*

H_1 is *that despite the fact that the volume of traffic has decreased in the time interval on three of the roads, the increase in vehicles per hour on the others is so great that the increased volume is significant overall.*

Rejection level is decided at $\alpha = \cdot 05$.

Because the data consist of matched pairs the Wilcoxon test is used to determine the significance of the variations in traffic flow.

Table 4.6 *Vehicles per hour at ten checkpoints this year and five years ago*

Checkpoint	1 5 years ago X	2 This year Y	3 d X–Y	4 Rank	5 R_1 +	6 R_2 –
1	101	132	–31	6		6
2	93	89	+4	1	1	
3	220	247	–27	5		5
4	266	255	+11	2	2	
5	287	325	–38	7		7
6	243	297	–54	9		9
7	234	301	–67	10		10
8	273	248	+25	3·5	3·5	
9	172	211	–39	8		8
10	199	224	–25	3·5		3·5
					6·5	48·5

To carry out the test the matched pairs are best set out as in Table 4.6, columns 1 and 2. Then for each pair the figure in column 2 is subtracted from column 1, and the result (d) placed in column 3, *noting the sign.* The ds (differences) are then ranked (column 4) *irrespective of sign.* Place the rank of each positive d in column 5, and the rank of each negative d in column 6, and sum individually columns 5 and 6. The smaller of the two sums is termed T. In this case the smaller is column 5, and therefore $T = 6{\cdot}5$.

Tied ranks are dealt with in the usual way by allotting ties the average of the ranks they would otherwise have occupied had there been slight differences between the values. Thus –25 and +25 occupy ranks 3 and 4, and each is allotted the rank of 3·5. Tied ranks have little effect on the result of the test.

Should two values in one pair be exactly the same (i.e. when the difference (d) between the two values in one pair is 0), that pair is ignored in the calculation of T, and n, the number of pairs in the sample, is reduced by one.

If the values of the paired numbers are randomly distributed, the differences between them will also be randomly distributed, together with their signs. So when the ranks are divided according to sign, the sums of columns 5 and 6 will be roughly equal. The smaller the sum of either column, the greater will be the difference between the values of X and Y. The smaller sum, T, is thus a measure of this difference.

Appendix 5 gives the significance of different values of T from $n = 6$ pairs to $n = 25$ pairs. It will be seen that with $n = 10$ the largest permitted value of T to be significant at the ·05 level is $T = 8$. With $n = 10$ our calculated value of T was 6·5. We are thus able to reject H_0 in favour of H_1, and state that the differences in traffic volume are significant at the ·05 level.

When the number of pairs exceeds 25, Appendix 5 cannot be used to establish significance. However, when this happens T may be found in the way described above, and then converted to a z-score, using Formula 4.6, to determine probability.

$$z = \frac{T - \dfrac{n(n+1)}{4}}{\sqrt{\dfrac{n(n+1)(2n+1)}{24}}}$$

Formula 4.6

Table 4.7 shows the yield of wheat in kilogram per hectare in selected parishes in England for 1927 and 1971.

The null hypothesis, H_0, *is that there is no difference in terms of wheat yield between 1927 and 1971.*

H_1 is *that there is a significant difference in yield between the two dates.*

Rejection level is decided at $\alpha = {\cdot}05$.

Ties are calculated as before and a z-score is obtained by substituting in Formula 4.6. Note that the number of pairs in the sample is reduced by one because one pair of values is the same, and so has been discounted.

$$z = \frac{T - \dfrac{n(n+1)}{4}}{\sqrt{\dfrac{n(n+1)(2n+1)}{24}}}$$

$$= \frac{96{\cdot}5 - \dfrac{30(30+1)}{4}}{\sqrt{\dfrac{30(30+1)(2 \times 30+1)}{24}}} = 2{\cdot}8$$

Table 4.7 *Wheat yield in kg/ha for 31 parishes for the years 1927 and 1971*

X	Y	X−Y	Ranks of X−Y	Ranks of positive values
658	675	−17	16	
631	656	−25	23·5	
582	578	+4	2·5	2·5
600	600		Discounted	
670	687	−17	16	
685	701	−16	16	
740	747	−7	5	
736	751	−15	12·5	
603	590	+13	10	10
594	588	+6	5	5
611	590	+21	19	19
625	650	−25	23·5	
675	709	−34	27	
656	679	−23	20·5	
640	652	−12	10	
670	677	−7	7	
732	728	+4	2·5	2·5
693	718	−25	23·5	
683	660	+23	20·5	20·5
668	677	−9	8	
627	611	+16	16	16
654	666	−12	10	
660	662	−2	1	
672	697	−25	23·5	
648	663	−15	12·5	
631	638	+7	5	5
638	622	+16	16	16
598	625	−27	26	
644	697	−53	28	
617	675	−58	30	
590	646	−56	29	
				$T = \overline{96{\cdot}5}$

It can be seen from Appendix 1 that a z-score of 2·8 has an associated probability of ·005. It is therefore possible to reject H_0 at the ·01 level of significance, and to accept H_1, and we may conclude that there was a significant increase in wheat yield between the two years tested.

Exercises

1. The mean wheat yield in kg/ha for the Welsh counties is given below (years 1966–70).

Anglesey	672	Glamorgan	668
Breconshire	590	Merioneth	604
Caernarvonshire	625	Monmouthshire	582
Cardiganshire	609	Montgomeryshire	677
Camarthenshire	623	Pembrokeshire	636
Denbighshire	624	Radnorshire	621
Flintshire	603		

State your hypotheses and test whether the mean wheat yield in Wales is significantly different from that in a) southern England b) northern England. (Use the mean values for England from Table 4.2).

At what level of significance is your result?

2. During the geomorphological study of two hillsides, termed areas A and B, with the same aspect but composed of different rock types, it is decided to ascertain whether the mean angles of slope developed on the rock in area A differ significantly from those in area B.

It is possible to make 21 measurements of slope at randomly selected locations in area A; but only 12 measurements are possible in area B. In each case slope measurements of the whole profile are made, and the mean of these measurements taken as the slope angle at that location.

For the purposes of this study the slope is deemed to terminate at the place where the angle of the profile falls below 4·0.

The following data are recorded:

Area A 7·9, 7·1, 6·0, 5·4, 4·9, 4·2, 6·0, 6·4, 4·9, 8·2, 6·7, 7·5, 5·5, 9·3, 9·7, 11·3, 6·2, 12·9, 8·5, 9·5, 12·2

Area B 9·6, 13·7, 13·6, 9·9, 8·9, 8·3, 4·5, 13·2, 4·9, 12·3, 12·5, 11·4.

(a) What are the means of the slope angles in area A, and area B?
(b) What is the percentage difference between them?
(c) State your hypotheses and test whether the difference is significant and at what level.
(d) What is the probability that differences of slope angle of this magnitude might have occurred through random (chance) variations?

3. During the survey of a pebble beach three distinct ridges are observed. The highest is the storm beach, consisting of pebbles thrown up by storm

waves at high water. The next ridge is that formed at high water by spring tides, and the lowest ridge at high water by neap tides. It appears that there is some relationship between the three ridges and the size of pebbles of which they are made (for the purpose of this study, 'size' may be determined by weight). To test the relationship, random samples of equal weight are taken from each ridge. The pebbles are then weighed individually and their overall weight obtained. They are then assigned to two categories, those heavier than the mean are called 'large', and those lighter than the mean 'small'.

The following numbers of pebbles are counted:

	Large pebbles	Small pebbles
Ridge on storm beach	10	22
Ridge at spring tide	29	20
Ridge at neap tide	16	21

Formulate your hypotheses and test whether there is a significant difference in terms of pebble size between the three beach ridges.

4. During the course of a study of a relatively flat area of agricultural land covered with varied deposits of Pleistocene age, it is decided to investigate the distribution of pasture from north to south by means of a transect. Every farm along the line of transect is visited and classified according to whether there is more, or less, than 50% of the holding under grass. The former are classed as 'pastoral' farms (P), the latter as 'not pastoral' farms (NP). The resulting data are as follows:

Type of farm P P P P P NP P NP NP NP NP NP NP NP NP P NP P NP P P P P

North ←—line of transect—→ South

The transect suggests a considerable difference in pastoral farming between the central part of the area, and the northern and southern extremities.

Formulate hypotheses and test whether this difference may be considered significant.

5. Divide the annual precipitation totals for Cropston given in Chapter 2, Exercise 5 (p. 26), into two categories, those years with precipitation above the mean, and those below. Use the runs test (as p. 57) to determine whether sequences of years when the rainfall was above or below the mean over the period 1872–1966 is statistically significant.

6. As part of the study of a large city it is decided to try to find out objectively (i.e. without asking people) whether there is any variation in the numbers of shoppers in the central area on two week-days, Monday and

Thursday. Ten large shops are selected and the number of people entering them on both days during shopping hours is recorded. The number of people counted on both days for each shop is as follows:

Shop	A	B	C	D	E	F	G	H	I	J
Monday	2740	2217	2385	3048	1304	3252	2571	2932	3419	2492
Thursday	2406	1920	2300	3124	987	3005	2431	2721	3472	2602

What is the difference in the number of people using the ten shops on Monday and Thursday?

State your hypotheses and test whether the difference is significant, and if so, at what level.

Why is it not valid to use a Mann Whitney U-test in this example?

7. At a time of fluctuating retail prices it is decided to check the changes over an interval of one year. The prices of 34 common user items are recorded. One year later the prices of similar items in the same retail stores are again observed. The prices in new pence for both occasions are as follows:

Prices one year ago	103	89	47	39	52	75	113	19	35	51	25	
Present price	130	115	60	31	49	68	125	18	41	43	30	
Prices one year ago	81	58	16	29	48	28	92	134	93	82	106	
Present price	92	70	14	33	54	28	103	151	93	94	96	
Prices one year ago	73	57	31	42	68	87	79	17	39	63	92	17
Present price	70	51	31	45	76	93	79	15	51	71	85	20

Most prices have gone up; a few have remained the same; and a few have gone down.

(a) What is the probability that changes in price are due to random variations?

(b) State your hypotheses.

(c) Are you able to reject H_0? If so, at what level of significance?

Chapter 5

Estimates from samples

The sampling distribution

Sampling techniques have been dealt with at some length in S.I.G. 2, *Data Collection.* Sampling will not be discussed here except to emphasize two most important points, which are essential if estimates made from samples, as described in this chapter, are to be valid. Firstly, the sample must be **random**. That is, it must be free from bias. Secondly, the observations must be **independent**. By 'independent' we mean that every measurement must have an equal probability of being included in the sample, and the selection of one must not influence the selection of another.

Suppose we take a random sample of 35 pebbles from a beach and calculate the roundness coefficient for each. To what extent can we be sure that the sample represents, in terms of roundness, the whole population (i.e. all the pebbles on the beach) from which it is drawn? To put this another way, it is very unlikely that the mean of our sample will be precisely the same as the mean of the whole population. How then can we estimate the extent of the error which is likely to be involved? To answer this question let us refer to the actual values in the sample contained in Table 5.1. (The way the values are distributed is shown in Fig. 5.1)

Table 5.1 *Roundness coefficients of a random sample of 35 beach pebbles*

35	36	32	29	28	30	58	34	37	33	36	36
35	31	27	42	21	27	29	34	38	34	39	40
31	42	46	29	19	29	35	39	41	43	30	

Suppose we took a large number of samples from the beach, we should find it very rare that any two samples had exactly the same mean or standard deviation. Most of the sample means would not differ by very much one from another, but sometimes, due to chance variations in the random

Fig. 5.1 *Frequency distribution, in class intervals of 4, for the values contained in the sample in Table 5.1*

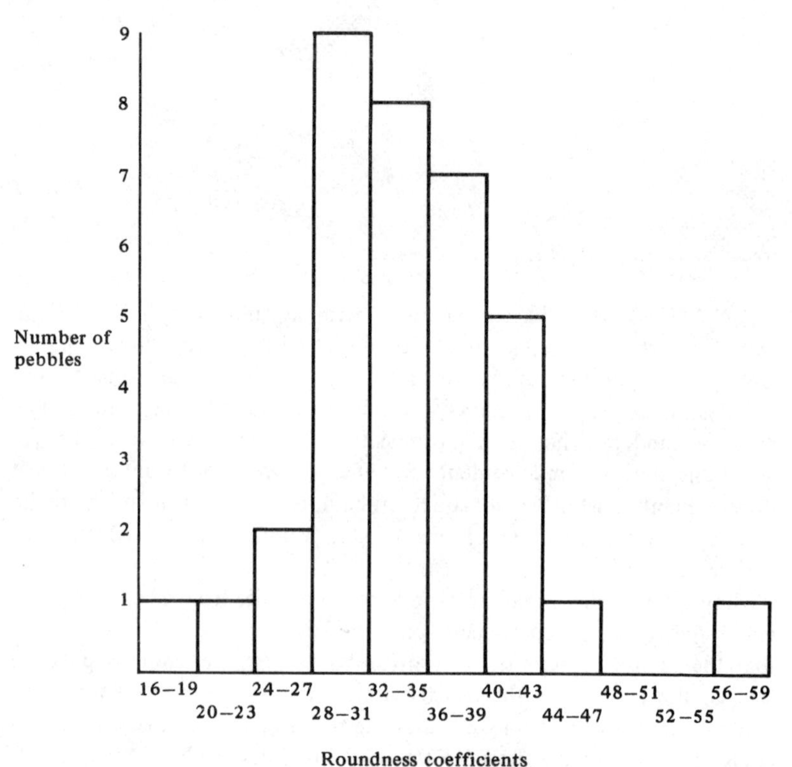

sampling method, we would obtain samples with mean values considerably greater or smaller than the majority. The distribution of all these mean values is known as the **sampling distribution.**

The standard error of the mean

It can be shown that if we took all possible samples of similar size from our population of beach pebbles, the sample means would form a *normal distribution* with the same mean as the parent population, *provided that the size of each of the samples is more than 30.* This holds true, *whatever the distribution of the parent population.* In the case of our pebble sample, Fig. 5.1 provides evidence that, if we could count them all, the roundness coefficients of the whole population are probably *not* normally distributed. Nevertheless, if we took sufficient samples of the same size from the beach we should find that their *means* would closely approximate to the normal curve.

It can also be shown that the standard deviation of this sampling distribution, which is called the *standard error of the mean*, is found by the formula:

$$S.E. = \frac{\sigma}{\sqrt{N}}$$

Formula 5.1a

where σ is the standard deviation of the population and N is the size of sample.

Note that the standard deviation of the sampling distribution $[\sigma/\sqrt{N}]$ uses the standard deviation of the *parent population.* When we calculate a standard error, we are forced to use the standard deviation, s, of the *sample* we have obtained. This introduces an unavoidable margin of error. However, with samples over 30 the error is probably fairly small and in any case is divided by $\sqrt{N}$ to reduce it further. Thus in practice, the standard error is found by the formula:

$$S.E. = \frac{s}{\sqrt{N}}$$

Formula 5.1b

Our pebble sample has a standard deviation, $s = 7{\cdot}2$, and a mean, $\bar{x} = 34{\cdot}4$. Now we have seen (p. 30) that with a normal distribution, 68% of values lie within one standard deviation either side of the mean, 95% of values lie within two standard deviations of the mean, and 99% of values lie within three standard deviations of the mean. Therefore it is possible to say that there is a 95% probability that the mean of the sample we have taken, $\bar{x} = 34{\cdot}4$, lies within two standard deviations of the population mean, and a 99% probability that $\bar{x} = 34{\cdot}4$ lies within three standard deviations of the population mean.

This is the same thing as saying that there is a 95% probability the true mean (i.e. of the whole population) will be within two standard deviations of 34·4 or a 99% probability that it will be within three standard deviations of 34·4.

We have already seen that the standard deviation of sample means, or standard error, is given by Formula 5.1b, thus the standard error in our example is:

$$S.E. = \frac{s}{\sqrt{N}} = \frac{7{\cdot}2}{\sqrt{35}} = 1{\cdot}2$$

Therefore, from the evidence provided by the sample, we are able to say:
1. with 95% confidence that the parent population mean is between 34·4 ± 2 S.E.—between 34·4 + 2·4 and 34·4 − 2·4—that is, *between 36.8 and 32·0;*
2. with 99% confidence that the parent population mean is between 34·4 ± 3 S.E.—or between 34·4 + 3·6 and 34·4 − 3·6—that is, *between 38·0 and 30·8.*

These upper and lower values are generally termed the *95% and 99% confidence limits.*

Bessel's correction

There is a correction which is sometimes used to modify the standard deviation (s) of a sample so that it is a closer estimate of the standard deviation of the population (σ) from which it was drawn. This is known as Bessel's correction and involves multiplying the sample standard deviation, s, by $\sqrt{[N/(N-1)]}$. This 'best estimate' is generally written as $\hat{\sigma}$.

$$\hat{\sigma} = s \times \sqrt{\frac{N}{N-1}}$$

Formula 5.2

With samples larger than 30, $\sqrt{[N/(N-1)]}$ closely approximates to 1, and thus s is very little altered. In our pebble example the standard deviation, correct to the first place of decimals, is unaffected by Bessel's correction, and both $s = 7{\cdot}2$ and $\hat{\sigma} = 7{\cdot}2$ are true. Normally when a sample is over 30 Bessel's correction is ignored. But it is necessary to be aware of its existence, even when dealing with large samples, because it is normally incorporated in any pre-programmed statistical calculator, and if allowance is not made for the correction small puzzling differences may occur between standard deviations calculated from the same data by different methods.

It is important to remember that what is said in this chapter is only valid for samples over 30. (There are methods for making estimates from smaller samples by using the 't' distribution, but this is a statistic requiring the assumption of normal distribution of the parent population, and is not considered here.) The larger the sample the more accurate will be the estimate of the population mean. Suppose instead of a sample of 35 we had measured 100 pebbles, and found them to have the same mean and standard deviation as the example given above. This time the standard error would be:

$$S.E. = \frac{s}{\sqrt{N}} = \frac{7{\cdot}2}{\sqrt{100}} = \frac{7{\cdot}2}{10} = 0{\cdot}72$$

By taking a larger sample the standard error is reduced from 1·2 to 0·72, and the 95% confidence limits are reduced to 35·8 and 33, instead of 36·8 and 32·0.

The binomial standard error

The binomial standard error, for the geographer, is a very important characteristic of the **binomial distribution**, because it provides a way of making estimates from samples which are in the form of percentages. So far, when making estimates from samples, we have considered data which have been *measured* on a continuous scale. Frequently the geographer's data is not measured by weight, or angle, or on a linear scale, but takes the form of a **count.**

Space does not permit a discussion of the characteristics of the binomial distribution in this book, except to say that **it deals with discrete data, and is applicable in situations of an 'either/or' nature.** (Further information concerning the characteristics of the binomial distribution can be found in *Quantitative Techniques in Geography: An Introduction* by Robert Hammond and Patrick McCullagh [O.U.P., 1974].) If we are studying the number of steel workers in different areas, we are dealing with discrete data: you can't have 1·5 steel workers, other than as a ratio. We are dealing with data that is *counted* (the number of workers) and we have a binomial situation because each worker *either* is, *or* is not, a steel worker, and our count of workers in every town will show the proportion in each. It is when data is provided in this form that the **binomial standard error** can be used.

Suppose we wish to establish the influence of a town upon the surrounding rural area, and choose as one of our criteria the areas of origination of non-urban dwellers who do their weekly shopping in the town on Saturday. As it is physically impossible to question everyone (and if it were some people might not be prepared to answer) it is decided to ask a random sample of 100 country dwellers. (Any town dwellers inadvertently questioned, or those who refuse to answer, would not count against the sample total.) The countryside surrounding the town is divided up into five areas, A, B, C, D, and E, and individuals are simply asked in which area they live. The result is contained in Table 5.2.

Table 5.2 *Number of Saturday shoppers originating from five previously defined areas.* $N = 100$.

A	B	C	D	E
25	31	20	7	17

The problem is to establish confidence limits for these sample figures. Fortunately, the binomial standard error is available for such a case. This is found by the formula:

$$S.E. = \sqrt{\frac{p\%.q\%}{N}}$$

Formula 5.3

where p is the percentage in one category, q is the percentage in the other category, and N is the number in the sample.

It will be seen that 25 people came from area A. As the sample was of 100 we can say 25% of the sample came from area A. If 25% came from A then 75% did *not* come from that area. Here we have our two categories, *either p% or q%.* Applying Formula 5.3 we find standard error for area A:

$$S.E._A = \sqrt{\frac{p\%.q\%}{N}} = \sqrt{\frac{25 \times 75}{100}}$$
$$= \sqrt{18 \cdot 75}$$
$$= 4 \cdot 3\%$$

To calculate the 95% confidence limits for the sample from area A we need to take two standard errors either side of the observed sample value. In this case S.E. = 4·3, so 2 × S.E. = 2 × 4·3 = 8·6. Therefore the 95% confidence limits for the sample from area A are:

$$25\% + 8 \cdot 6\% = 33 \cdot 6\%$$
$$\text{and} \quad 25\% - 8 \cdot 6\% = 16 \cdot 4\%$$

In other words we can say with 95% confidence that between 16·4% and 33·6% of the whole non-urban dwelling population of shoppers who were in town that Saturday lived in area A.

Similarly the 95% confidence limits for the sample from area D:

$$S.E._D = \sqrt{\frac{p\%.q\%}{N}} = \sqrt{\frac{7 \times 93}{100}}$$
$$= \sqrt{6 \cdot 51}$$
$$= 2 \cdot 5\%$$

The 95% confidence limits are therefore 7% ± (2 × 2·5)% or between 12% and 2%.

These confidence limits are very wide. Indeed they may be unacceptably wide. To reduce them it is necessary to increase the size of the sample. (Notice it is the size of the *sample* that matters, not the size of the *population* from which it is drawn). If the sample size in the above example were increased to 500 and if the same *proportions* of people originated from areas A to E, the standard error for the 25% from area A becomes:

$$S.E._A = \sqrt{\frac{p\%.q\%}{N}} = \sqrt{\frac{25 \times 75}{500}}$$
$$= \sqrt{3{\cdot}75}$$
$$= 1{\cdot}94\%$$

and the 95% confidence limits are reduced from between 33·6% and 16·4% to 25% ± 2 × 1·94, or between 28·9% and 21·1%.

Similarly the standard error for the 7% from area D becomes:

$$S.E._D = \sqrt{\frac{p\%.q\%}{N}} = \sqrt{\frac{7 \times 93}{500}}$$
$$= \sqrt{\frac{651}{500}}$$
$$= 1{\cdot}14\%$$

and the 95% confidence limits are reduced from between 12% and 2% to 7% ± 2 × 1·14, or between 9·3% and 4·7%.

Notes 1. The results of a sample survey of this kind are very often presented in the form of a proportional flow-line diagram. When this is done it is informative to indicate *on the diagram* the extent to which the width of the flow-line is increased by the 95% or 99% confidence limits.

2. This technique is very useful for establishing the S.E. of a random point sample taken from a map. For example, if a sample of arable land is taken from a land use map using a super-imposed grid with randomly selected coordinates, *p*% of the points will fall on the arable land (representing the percentage of the area which is under arable cultivation), and *q*% will not. Confidence limits for this sample percentage can be established using the binomial S.E.

Estimating sample size

The influence that sample size has over confidence limits is illustrated above, and it is common sense that the larger the sample taken the greater is the probability that it accurately reflects the distribution from which it was drawn: as size increases confidence limits get closer to the sample mean. The trouble is that taking large samples may be very time-consuming, or may be very costly. For example, opinion polls on the way the electorate say they are going to vote can become very popular as a general election approaches. In evaluating their worth it is necessary to take into consideration their standard error. If a sample of 1200 is taken (a size that has been used for this purpose) and the result shows that 52% intend to vote Labour and 48% Conservative:

$$S.E. = \sqrt{\frac{p\%.q\%}{N}} = \sqrt{\frac{48 \times 52}{1200}} \simeq 1{\cdot}5\%$$

The 95% confidence limits therefore for Labour are 49% to 55% and for Conservative are 45% to 51%. On this basis the opinon poll could be interpreted statistically as anything between a marginal Conservative victory, a landslide victory for Labour, or a dead heat!

Whether we are conducting a shopping survey, forecasting an election result, or wondering how large a sample will be necessary to give a reasonably accurate estimate of the mean size of pebbles at a particular place on the heaped masses of the Chesil Bank, it is necessary first to have some idea of how large the sample must be to give the degree of accuracy we desire. A larger sample than this will be a waste of money, or time. Fortunately, the size of sample required is easily estimated using the standard error formula.

The first step is to take a pilot sample, which should be more than 30. If the data are in the form of *proportions* then the binomial S.E. is used. We decide the largest standard error that is acceptable for our purposes.

Let the required S.E. = d.

Then $$d = S.E. = \sqrt{\frac{p\%.q\%}{N}}$$

or $$d^2 = \frac{p\%.q\%}{N}$$

and $$N = \frac{p\%.q\%}{d^2}$$

Formula 5.4

Let us suppose we are conducting a transport survey to find out the number of shoppers using public transport, and are prepared to accept S.E. = 2%. The pilot survey is of 50 people and reveals that 30 of them (60%) used public transport. Then the size required (N) to give a sample S.E. of 2 is:

$$N = \frac{p\%.q\%}{d^2} = \frac{60 \times 40}{2^2}$$

$$= \frac{2400}{4} = 600$$

If we are dealing with *measured* data it will be remembered that the standard error is found from S.E. = $\sigma/\sqrt{N}$. Therefore if d is the acceptable S.E.:

$$d = \frac{\sigma}{\sqrt{N}}$$

or

$$\sqrt{N} = \frac{\sigma}{d}$$

or

$$N = \left(\frac{\sigma}{d}\right)^2$$

Formula 5.5

It is apparent from Formula 5.5 that the size of sample depends partly upon the confidence limits considered acceptable, and partly upon the standard deviation of the pilot sample. The distribution of values in the sample shown in Fig. 5.1 indicates values clustering fairly close to the mean, and the sample has a relatively small standard deviation of 7·2, with a standard error of 1·2. If it is wished to reduce this standard error to 1, using Formula 5.5, the size of sample required is:

$$N = \left(\frac{7{\cdot}2}{1}\right)^2$$
$$= 7{\cdot}2^2$$
$$= 52$$

However, had the standard deviation been twice as great, to obtain the same S.E. = 1, we find:

$$N = \left(\frac{14{\cdot}4}{1}\right)^2$$
$$= 14{\cdot}4^2$$
$$= 208$$

Had the standard deviation been twice as great the size of the sample would have had to be increased four times to achieve the same standard error.

Exercises

1. From a random sample 64 farms are found to have a mean area of 45 hectares, with a standard deviation of 12.

(a) What are the 95% and 99% confidence limits for the mean area?

(b) What effect would there be on the confidence limits if the size of the sample were increased to 144?

2. 40 random soil samples were taken from the surface of a gravel terrace. By drying and sieving each sample the gravel content was removed and then weighed. The mean weight of gravel over all samples was found to be 920 gm, with a standard deviation of 90.

(a) What are the 95% and 99% confidence limits for the mean gravel content of the soil?

(b) What exactly do these confidence limits tell you?

(c) What would be the effect on the confidence limits if the sample size were increased to 70?

3. It is desired to construct a flow line diagram between town A and town B to show the proportion of visitors to A who make the journey from B by road, and the proportion who travel by rail. To this end a random sample of 100 visitors from B is taken in the shopping centre of A. 40 people are found to have arrived by rail and the remainder by road. It is decided to draw the flow line at a scale of 1 mm = 2%. The width of the flow line for those arriving by rail would therefore be 20 mm, and those travelling by road 30 mm. By how much would the width of these lines need to be increased to indicate:

(a) the 95% confidence limits?

(b) the 99% confidence limits?

4. In a 10% sample of a town with a working population of 20 000, it is found that 8000 are employed in service industries. What are the 99% confidence limits for the number employed in service industries in the town?

5. A random sample of 100 pebbles is taken from a moraine.

(a) The roundness coefficients are measured and found to have a standard deviation of 15. What size of sample is required if the standard error is to be no larger than 1?

(b) 30 of the pebbles are found to be of flint. What size of sample would be necessary:

(i) for the S.E. to be not greater than 3?

(ii) for the 99% confidence limits to be not more than ± 6%.

Chapter 6

Correlation

The meaning of correlation

It is sometimes said of Britain that there is an areal **correlation** between low incomes and high unemployment. This statement means that in those areas where unemployment is high, incomes tend to be low, and that people in areas with a lower percentage of unemployed persons tend to have higher per capita incomes. In other words, **there is a mathematical *association* between the two sets of values,** unemployment, and per capita income. Notice that in this case the data is *paired*. That is, we are relating the unemployment figure and per capita income for the *same area* in each case. **Correlation can only apply when paired data are used.** Pairing may be achieved in several ways. For example, wheat yield per acre last year for a series of parishes may be compared with wheat yield per acre 100 years earlier *for the same parishes.* Or it is possible to correlate climatic statistics, say mean annual rainfall, with the altitude of each station in a sample, so that the mean annual rainfall, and the altitude of each station, provide a matched pair of values.

A **correlation coefficient*** is a mathematical measure of the degree of association between two paired variables. (The symbol representing the coefficient depends upon the method used. Two formulae are used in this chapter, Kendall's represented by τ, and Spearman's represented by r_s.) The correlation formula is so arranged that when one variable *increases* in exact proportion to the other, the coefficient will be +1·0. If one variable *decreases* in exact proportion to the other the coefficient will be −1·0. When there is *no* association between the two variables the coefficient will be near 0. (It will rarely be exactly 0, because of random variations within the sample.) Consider two variables A and B:

A	1	2	3	4	5
B	2	4	6	8	10

*The importance of correlation in the application of the scientific method to geographical problems was discussed in S.I.G. 1, Chapter 1.

Fig. 6.1 *Graphic representation of correlation*
A. Perfect positive correlation. One variable increases in direct proportion to the other.
B. Perfect negative correlation. As one variable increases the other variable decreases in direct proportion.
C. Twenty-five pairs of numbers taken from a table of random numbers, disclosing no association, and representing a random distribution

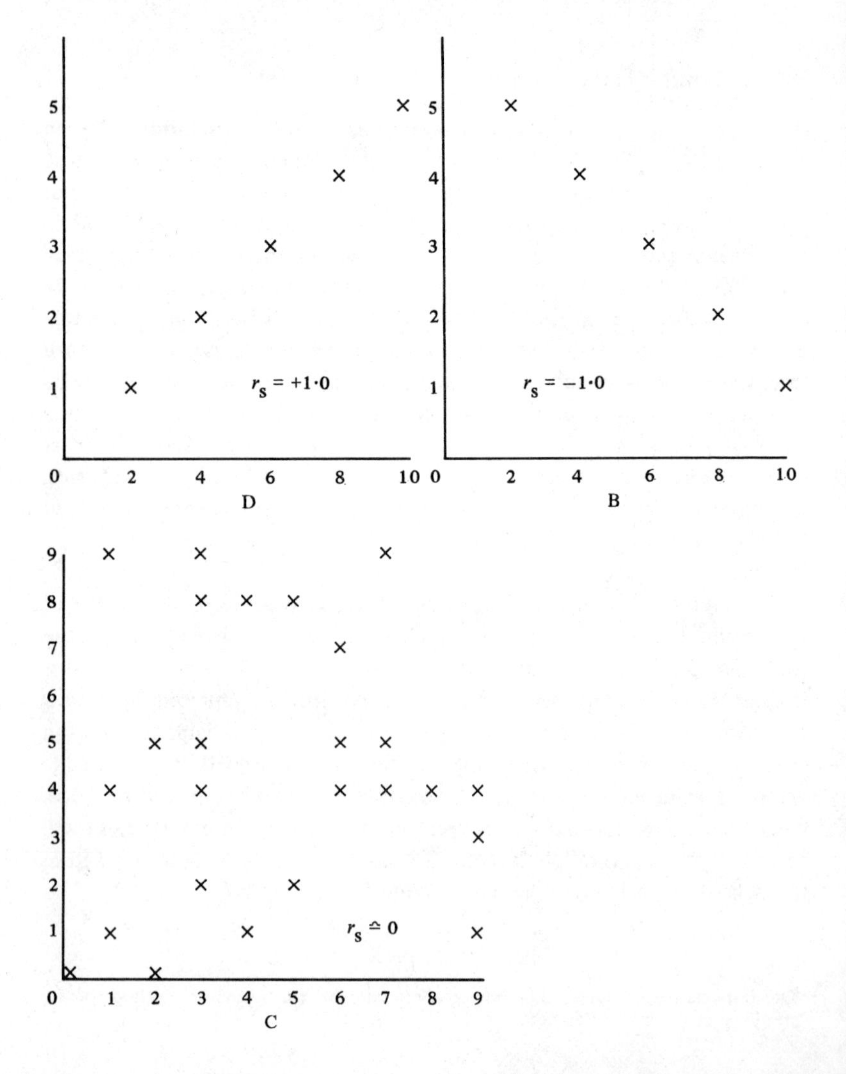

These are said to have a 'perfect positive correlation of +1·0'. Similarly if the order of B were reversed we should have a 'perfect negative correlation of −1·0'.

A	1	2	3	4	5
B	10	8	6	4	2

The implications of both cases are clearly shown in Fig. 6.1, together with twenty-five paired values randomly distributed.

It will be seen that when $r_s = +1{\cdot}0$, represented in Fig. 6.1A, the points lie in a straight line rising from left to right. When $r_s = -1{\cdot}0$, represented in Fig. 6.1B, the points are also in a straight line but the slope is in the opposite direction. The random distribution shown in Fig. 6.1C has no obvious pattern, and discloses no association between the two variables.

The correlation coefficient is a number which *describes* the degree of association between two variables, and as such it is a useful method of comparison. But like all statistical tests, it is also necessary to calculate the probability that the apparent association may be the result of random variations, that is, we must also test whether the coefficient is *significant*. Two methods of calculating and testing a correlation coefficient are given below, but first it is important to emphasize most strongly that a **correlation coefficient, shown to be statistically significant, does not necessarily imply that there is a causal relationship between the two variables,** because the mathematical association observed may have been caused by a third factor. For example, the abundance of a moorland plant species may be found to correlate significantly with certain altitudes. This does *not* necessarily mean that *altitude is the causal factor*. As found in some of the limestone areas in Yorkshire, although a significant correlation exists between altitude and moorland vegetation, the *causal* factor which really affects the distribution is the development of an acid type of soil on residual patches of gravel deposited as a result of a Pleistocene glaciation. Statistical techniques are important and useful tools, but the correct interpretation of the results rests upon the knowledge and judgment of the observer. This is particularly important in evaluating correlations.

Kendall's correlation coefficient τ

For small samples

Kendall's τ (tau) is a particularly useful method of calculating a correlation coefficient because:

(a) it can be used safely with small samples ($N \leqslant 10$);

(b) it is distribution free and no assumption need be made concerning the background population;
(c) it is quick and easy to calculate;
(d) for samples where $N > 10$ it is possible to calculate probability from a z-score by means of a very simple formula.

The method of calculating τ is best shown by an example. Suppose we had two variables, X and Y, consisting of seven pairs. First, allot a rank to each value, ranking each sample in *separate* sequence.

Ranks of X	7	3	5	1	2	6	4
Ranks of Y	6	4	5	2	1	7	3

Then place the ranks of X in the normal sequence 1, 2 . . . 7, and place below each rank of X the appropriate paired rank of Y.

Ranks of X	1	2	3	4	5	6	7
Ranks of Y	2	1	4	3	5	7	6

Observe the ranks of Y, beginning from the left (in this case) with the rank of 2. Allot +1 for each rank of Y *to the right* that is greater than 2, and −1 to each rank of Y to the right which is less than 2. The sum of these plus and minus values is the score for the left hand rank (2) of Y. Thus the total for the left hand rank (2) of Y is −1 +1 +1 +1 +1 +1 = +4. Carry out the same procedure for each rank of Y in turn. Add together the score for each rank of Y to form a total score termed 'S'.

Ranks of Y		Score
2	(−1) + (+5)	= +4
1	+5	= +5
4	(+3) + (−1)	= +2
3	+3	= +3
5	+2	= +2
7	−1	= −1
		S = +15

The maximum possible value for S for any series of numbers in *normal* sequence, i.e. 1, 2, 3 . . . N, may be found by the formula $\frac{1}{2}N(N-1)$, where N is the number of values in the sequence. Thus the score for X with the ranks placed in normal order is:

$$(\tfrac{1}{2})(7)(7-1) = \frac{42}{2} = 21$$

This is the same figure as that obtained by adding up the individual scores of the X ranks as we did for Y.

Ranks of X	Score
1	+6
2	+5
3	+4
4	+3
5	+2
6	+1
	+21

The correlation coefficient τ is found by dividing the observed value of S for a given sequence by the maximum value obtained if the ranks were re-arranged in normal order $(1, 2, 3 \ldots N)$. Thus the formula for finding τ is:

$$\tau = \frac{S}{\tfrac{1}{2}N(N-1)}$$

Formula 6.1

We have already found the value of $S = 15$ in the example above with $N = 7$, and can now substitute these values in Formula 6.1 to find τ

$$\tau = \frac{15}{\tfrac{1}{2}(7)(7-1)}$$

$$= \frac{15}{21}$$

$$= +0{\cdot}71$$

The correlation coefficient therefore is $\tau = +0{\cdot}71$, which is a mathematical *description* of the degree of association between variable X and variable Y. It still remains to establish its significance. This depends upon the number of pairs in the sample (N), and the value of S, and may be obtained from the table in Appendix 6. In our case $N = 7$ and $S = 15$.Appendix 6 shows that these values of N and S have an associated probability of ·015. We can, therefore, say that $\tau = 0{\cdot}71$ is significant at the ·05 level and very nearly significant at the ·01 level.

Table 6.1 shows the percentage unemployment rate in 1969 and 1971, and the average income per person in 1967–68, for each of the ten New Standard Regions of Britain. It has been suggested earlier that in Britain areas of high unemployment tend to be those of low per capita income. In times of relative industrial stagnation this association tends to become intensified. Table 6.1 appears to bear this out, but in order to make precise comparisons and to determine whether the result is statistically significant it is necessary to test the figures quoted. A correlation is well-suited to do this.

Table 6.1 *Unemployment and average income per person in the ten New Standard Regions of Britain*

Region	Percentage unemployment in 1969		Percentage unemployment in 1971		Average income per person (in £s) 1967–68	
	Z	Rank	Y	Rank	X	Rank
North	4·8	1	5·2	2	1087	10
Yorkshire and Humberside	2·6	5	3·6	4·5	1113	6
E. Midlands	2·0	7·5	3·1	8·5	1152	3
E. Anglia	1·9	9	3·1	8·5	1115	5
South East	1·6	10	2·0	10	1275	1
South West	2·7	4	3·5	6	1130	4
W. Midlands	2·0	7·5	3·2	7	1180	2
N. West	2·5	6	3·6	4·5	1103	8
Wales	4·1	2	4·6	3	1104	7
Scotland	3·7	3	5·7	1	1093	9

Let us first test for association the relationship between low personal income (X) and percentage unemployment in 1969 (Z).

Our null hypothesis, H_0, is that *there is no association between low personal income and high unemployment, and that the observed differences are the result of chance variations.*

The alternative hypothesis, H_1, is that *the highest rate of unemployment exists in areas of low personal income.*

An acceptable rejection level is decided at $\alpha = \cdot 05$.

To carry out the test the ranks of the income variable (X) are placed in normal sequence, and the corresponding ranks of the 1969 unemployment variable (Z) placed below. The scores of each rank of Z are then summed to find S. Ties are dealt with in the usual way. That is, equal values are allotted the average of the ranks they would have received if there had been slight differences between the values.

If the score is being calculated for a tied rank, all the ties will, of course, have the same rank value. Any tie of the same value to the *right* of the observed rank will contribute 0 *to the sum of that rank*. All ties may contribute to the scores of other rank values. In the example below, when considering the score of the left hand rank of Z (i.e. 10) each rank of 7·5 contributes −1. But when the score of the left hand rank of 7·5 is being considered, the tied rank of 7·5 to the right contributes 0.

X	1	2	3	4	5	6	7	8	9	10
Z	10	7·5	7·5	4	9	5	2	6	3	1

$$S = (-9) + (0 - 6 + 1) + (-6 + 1) + (-3 + 3) + (-5) + (-3 + 1) + (+2 - 1) + (-2) + (-1)$$
$$= (-9) + (-5) + (-5) + (0) + (-5) + (-2) + (+1) + (-2) + (-1)$$
$$= -28$$

We now find the value of τ using Formula 6.1.

$$\tau = \frac{S}{\frac{1}{2}N(N-1)}$$

where S is the sum of the scores of each rank of Z, and N is the number of pairs.

$$\tau = \frac{-28}{\frac{1}{2}(10)(10-1)}$$
$$= \frac{-28}{45}$$
$$= -0{\cdot}62$$

Reference to Appendix 6 shows that with $S = 27$ and $N = 10$ there is an associated probability of ·0083. That is, τ is significant at the ·01 level. We are therefore able to reject H_0, and assert with 99% confidence that in 1969 there was a significant spatial association between low personal income and areas of high unemployment in Britain.

But what of the situation in 1971? Unemployment had increased; but had it increased proportionately more or less in areas of relatively low personal income? The association of the previous variable, personal income (X), can again be tested with the new variable, percentage unemployment in 1971 (Y). The procedure is as before.

X =	1	2	3	4	5	6	7	8	9	10
Y =	10	7	8·5	6	8·5	4·5	3	4·5	1	2

$$S = (-9) + (+2 - 6) + (0 - 6) + (+1 - 5) + (-5) + (0 - 3) + (+1 - 2) + (-2) + (+1)$$
$$= (-9) + (-4) + (-6) + (-4) + (-5) + (-3) + (-1) + (-2) + (+1)$$
$$= -33$$

We find the value of τ using Formula 6.1.

$$\tau = \frac{S}{\frac{1}{2}N(N-1)}$$

where S is the sum of the scores of each rank of Y, and N is the number of pairs.

$$\tau = \frac{-33}{\frac{1}{2}(10)(10-1)}$$

$$= \frac{-33}{45}$$

$$= -0{\cdot}73$$

Reference to Appendix 6 shows that with $S = 33$ and $N = 10$, the associated probability (i.e., that such an association of paired values is due to chance) is ·0011. τ is thus significant at the ·01 level, meaning that there is less than 1 chance in 100 that the observed association of values in the two variables has come about through random variations.

We have now shown that in both 1969 and 1971 there existed a significant negative correlation between personal income and areas of relatively high unemployment. The increase in the value of the coefficient from –0·62 in 1969, to –0·73 in 1971, showed that the closeness of the association had *increased* by 11%. In other words we have a significant *descriptive* index not only of the association between the two variables, but also of the amount of change in the association between 1969 and 1971.

Fig. 6.2 *The average income per employed person in pounds sterling for 1967–68 and the percentage unemployment rate for 1969 and for March 1971, by New Standard Regions in Britain*

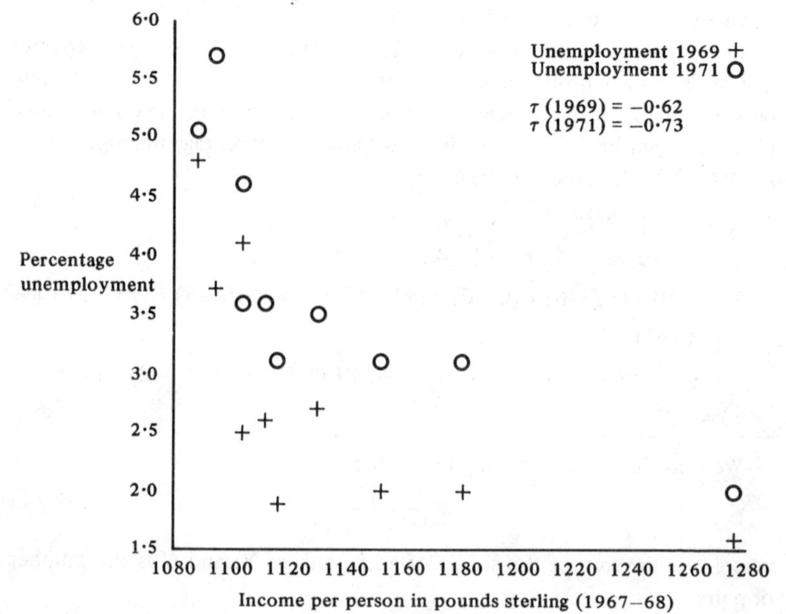

The situation is depicted graphically in Fig. 6.2. The points do not lie in a straight line, of course, because correlation is not perfect. But they are also obviously very far from being randomly distributed. The correlation coefficients are negative, and so the general trend on the graph is from high left to low right. It will be observed that the points for 1971 are generally higher up the graph (reflecting the bigger percentage of unemployed), and approximate more closely to a straight line (because τ is approaching more closely to a perfect correlation of $-1{\cdot}0$). The graph also shows well the uniqueness of the South East in terms of personal income and unemployment.

Kendall's correlation coefficient τ for large samples

When sample size is greater than 10, τ is calculated in the same way as demonstrated above, but its significance is tested by obtaining a z-score. The formula from which the z-score is derived is:

$$z = \frac{\tau}{\sqrt{\dfrac{2(2N+5)}{9N(N-1)}}}$$

Formula 6.2

Let us consider the data contained in Table 6.2

H_0 is that *there is no association between temperature range and latitude.*

H_1 is that *temperature range increases in proportion to distance from the equator,* i.e., temperature range increases as the angle of latitude increases both north and south of 0.

Let the level of rejection be $\alpha = {\cdot}01$.

We proceed as before, arranging the ranks of one variable (it does not matter which) in sequence, and placing the paired rank of the other variable beneath it.

X	1	2	3	4	5	6	7	8	9	10	11	12·5	12·5	14	15	16	17	18
Y	3	5	1·5	1·5	10	12	4	10	6	7	10	13	8	14·5	14·5	17	16	18

Note Tied values are dealt with as before, but the reader will observe a difficulty in this example, namely, that there are tied ranks in *both* X and Y. We have already seen above how to deal with the tied ranks of Y. But this time there are also two tied ranks in X corresponding to ranks 13 and 8 of Y. The problem is whether to place the paired ranks of Y in order 13,

Table 6.2 *Mean annual temperature range compared with latitude at 18 selected stations*

Station	Latitude (Y)	Rank of Y	Range (X)	Rank of X
Manaos	3S	18	3	18
Zanzibar	6S	17	6	16
Durban	30S	10	13	11
Caracas	10N	16	16	17
Mexico City	19N	14·5	11	14
Bombay	19N	14·5	10	15
Delhi	29N	12	34	6
Algiers	37N	6	25	9
Valparaiso	33S	8	12	12·5
Sydney	34S	7	19	10
Chunking	30N	10	38	5
Boston	42N	5	45	2
Paris	48N	4	28	7
Moscow	56N	3	54	1
McMurdo Sound	78S	1·5	40	4
Cairo	30N	10	27	8
Antofagasta	24S	13	12	12·5
Spitzbergen	78N	1·5	44	3

8, or 8, 13. In the first instance in calculating S when focussing on 13 the 8 would contribute -1, while in the second instance when focussing on 8 the 13 would contribute $+1$ to S. Therefore both ranks of Y paired with the tied rank of 12·5 in X are taken to contribute 0 to S *within the tie only*. They are counted as usual when compared with all other values.

$$S = (+15 - 2) + (-3 + 13) + (+14) + (+14) + (+7 - 4) + (-6 + 6) + (+11) + (-3 + 6) + (+9) + (+8) + (+6 - 1) + (+5) + (+5) + (+3) + (+3) + (+1 - 1) + (+1)$$

$$= 107$$

Therefore:

$$\tau = \frac{S}{\frac{1}{2}N(N-1)} = \frac{107}{\frac{1}{2}(18)(18-1)} = 0{\cdot}70$$

We are now able to use Formula 6.2 to obtain a z-score:

$$z = \frac{\tau}{\sqrt{\dfrac{2(2N+5)}{9N(N-1)}}}$$

$$= \frac{0{\cdot}70}{\sqrt{\dfrac{2 \times (2 \times 18 + 5)}{9 \times 18(18-1)}}}$$

$$= 4{\cdot}1$$

A score of 4·1 has an associated probability so small that it is not given in Appendix 1 (column B). We can therefore reject H_0 and accept H_1 at the highest level of significance.

Spearman's rank correlation coefficient

Spearman's correlation coefficient, frequently written r_s, has been used for many years, and is derived from Pearson's Product Moment formula, which is the most powerful correlation test available. However, the Product Moment coefficient, r, is a parametric statistic, and before it can be used the necessary requirements for any parametric test must be met (i.e. distribution of the population must be normal). Also unless a good calculating machine is available the arithmetic involved is laborious and time-consuming. (The method is explained in Appendix 8a.)

Spearman's alternative reduces the data to rank values. In doing so some of the information is lost, but the test is relieved of the assumptions necessary for Product Moment (i.e. a normal distribution of population need not be assumed, and only ordinal measurement is required), and the coefficient is very much quicker and simpler to calculate. r_s and τ are both equally powerful, having 91% of the power efficiency of Pearson's r. That is, r_s and τ with 100 paired values have the same efficiency in the rejection of H_0 as Pearson's r with 91 paired values.

The formula is a simple one:

$$r_s = 1 - \frac{6\Sigma d^2}{N^3 - N}$$

Formula 6.3

where d is the difference in rank of the values in each matched pair, and N is the number of pairs in the sample.

Table 6.3 *Per capita G.N.P. (U.S. $, 1968), and percentage population increase (1961–68) for north and central American countries with a population of more than 100 000 in 1968*

	1	2	3	4	5	6
Columns	Per Capita G.N.P.	R_1	Popn. Growth %	R_2	d	d^2
U.S.A.	3980	1	1·4	21·5	−20·5	420
Mexico	530	9	3·5	4·5	4·5	20
Canada	2460	2	1·9	18	−16	256
Cuba	310	17	2·4	13	4	16
Guatemala	320	16	3·1	9·5	6·5	42
Haiti	70	23	2·0	16	7	49
Dominican Rep.	290	18	3·6	2·5	15·5	240
El. Salvador	280	19	3·6	2·5	16·5	272
Puerto Rico	1340	4	1·8	19	−15	225
Honduras	260	20	3·4	6·5	13·5	182
Jamaica	460	11	2·0	16	−5	25
Nicaragua	370	15	3·4	6·5	8·5	72
Costa Rica	450	12	3·5	4·5	7·5	56
Panama	580	8	3·3	8	0	0
Trinidad and Tobago	870	6	2·6	12	−6	36
Martinique	610	7	2·0	16	−9	81
Guadeloupe	510	10	2·1	14	−4	16
Barbados	440	13	1·0	23	−10	100
Netherlands Antilles	1200	5	1·4	21·5	16·5	272
Bahama Islands	1460	3	5·8	1	2	4
Brit. Honduras	390	14	3·1	9·5	4·5	20
St. Lucia	220	21·5	2·9	11	10·5	110
Grenada	220	21·5	1·7	20	1·5	2
						$\Sigma d^2 = 2516$

Data Source: World Bank Atlas, 1970

There is considerable evidence today that rapid population growth is taking place in countries that are least able to support it economically. One way of measuring a country's wealth is by dividing the Gross National Product (the value of all the goods and services produced in the country in a year) by the total population. This reduces economic wealth to a ratio and thus makes one country directly comparable with another. Table 6.3

gives the per capita G.N.P. and percentage population growth rates for countries in north and central America with a population of more than 100 000.

Inspection of the data tends to support the view that population growth is inversely proportional to G.N.P. The U.S.A. with a per capita G.N.P. of 3980 has a population growth rate of only 1·4%, whereas the Dominican Republic and El Salvador, with a per capita G.N.P. of 290 and 280 respectively, both have a population growth rate of 3·6%. There are anomalies, however, and it is necessary to measure the extent of the association statistically. The two variables are in the form of matched pairs (we have per capita G.N.P. and percentage population growth for each country), so the calculation of a correlation coefficient would be a suitable test.

H_0 is *that no mathematical association exists between per capita G.N.P. and percentage population growth, and that observed differences are the result of random variations.*

H_1 is *that as countries become wealthier so the rate of population increase declines.*

The rejection level is decided as $\alpha = \cdot 05$.

To carry out the test the data have first to be ranked independently for each variable (ties are dealt with as previously). Ranks are shown in columns 2 and 4 on Table 6.3. The difference (d) in rank value for each pair is then recorded (column 5), and squared (column 6). Column 6 is then summed. There are 23 pairs in the sample, i.e., $N = 23$. It is now possible to obtain r_s from Formula 6.3.

$$r_s = 1 - \frac{6\Sigma d^2}{N^3 - N}$$

$$= 1 - \frac{6 \times 2516}{23^3 - 23}$$

$$= - \cdot 24$$

We now have a measure ($- \cdot 24$) which describes the association between the two variables. It shows a negative correlation, although a surprisingly low one. $r_s = \cdot 24$ can now be tested for significance using Appendix 7. It will be seen that for sample size $N = 22$ (even numbers only are shown) r_s must be at least ·36 to be significant at the ·05 level. We are therefore unable to reject H_0. It appears that, as far as our sample shows, there is NOT a very strong relationship between wealth and population increase. This may come as a surprise, but it should certainly cause us to examine the facts carefully.

Once the data is ranked, each country has exactly the same weight in the calculation regardless of size. Thus the U.S.A. (pop. 201 152 000) contributes the same to the final coefficient as Grenada (pop. 103 000). We also observe the Bahama Islands (pop. 177 000) with a per capita G.N.P. of 1460 and a population increase of 5·8%, the highest growth rate of all. When we were formulating our hypotheses we probably thought of 'population growth' as 'the natural increase of the indigenous population'. If so, to what extent is the population of the Bahama Islands due to 'natural increases', and how much to the immigration of wealthy, sun-seeking settlers? Should the Bahama Islands be excluded on these grounds?

It follows that, although statistical techniques help to bring precision to the evaluation of data, and rigour to thought processes, they do not eliminate the need for knowledge and judgment in the observer.

Comparing Spearman's r_s and Kendall's τ

It should be noted that the underlying scales of τ and r_s are different, and therefore the resulting coefficients are not directly comparable, although both have the same power efficiency. That is, using exactly the same data the numerical values of τ and r_s will be different. However, the significance of both will be the same, because both have equal power to reject H_0.

Kendall's τ is particularly useful for small samples. But when sample size greatly exceeds 20 the calculation of S becomes rather laborious and mistakes may possibly occur.

Spearman's r_s is easier to calculate and any mistake made is more obvious and more easily checked, but unfortunately there is some doubt about the significance table for r_s with sample sizes between 10 and 20. For this reason it is probably better to use τ unless the sample size is in excess of 20, when ease of calculation makes r_s more suitable. But remember that the actual coefficients obtained by using the two different methods will not be directly comparable numerically.

The chi-squared contingency coefficient C

The χ^2 test was used (pp. 6 and 49) to determine the statistical significance of frequency distribution comprising two or more categories. In this section the values of χ^2 previously obtained are used to calculate the

χ^2 contingency coefficient, a very useful and simple descriptive measure of association between two sets of data. Represented by C, it is a kind of correlation coefficient, and may be used when measurement is on a nominal scale for one, or both, of the attributes being studied, and the data set out in the form of a contingency table. It is found from the formula:

$$C = \sqrt{\frac{\chi^2}{\chi^2 + N}}$$

Formula 6.4

where N is the total number of observed frequencies and χ^2 is calculated from:

$$\chi^2 = \Sigma^r \Sigma^k \frac{(O - E)^2}{E}$$

Formula 4.4, p. 52

Two examples will demonstrate the use of C.

1. In Chapter 4, χ^2 was used to test the distribution of abandoned fields with relation to aspect (p. 52) from the information contained in Table 4.4a. A result was obtained of $\chi^2 = 13{\cdot}90$. This may be converted to a coefficient using Formula 6.4:

$$C = \sqrt{\frac{\chi^2}{\chi^2 + N}}$$
$$= \sqrt{\frac{13{\cdot}90}{13{\cdot}90 + 225}}$$
$$= 0{\cdot}24$$

0·24 is a measure of the association between the number of abandoned fields, and the direction in which they face, and it may be used as a tool for comparison with other areas. C can also be useful cartographically if it is desired to differentiate areas on a map where the degree of association between two attributes varies. (Remember that for any other correlation coefficient to be used, at least ordinal–ranked–data are required.)

2. Applying Formula 6.4 to the data contained in Table 4.4c (p. 53) we find:

$$C = \sqrt{\frac{\chi^2}{\chi^2 + N}}$$
$$= \sqrt{\frac{9{\cdot}05}{9{\cdot}05 + 108}}$$
$$= 0{\cdot}28$$

Normally it is necessary to test a coefficient for significance. That is, we test to estimate the probability that the values in our sample represent a chance association (which is not reflected in the population from which it was drawn). In the case of the contingency coefficient this is unnecessary, because the level of significance of C is the same as that for the value of χ^2. In 1. above, $\chi^2 = 13{\cdot}90$ did not reach our rejection level of $\alpha = {\cdot}05$, and the result of the C test could not be regarded as significant. $C = 0{\cdot}24$ therefore must remain only a description of the degree of association in our sample.

In example 2 however we have a slightly higher value, $C = 0{\cdot}28$, and in this case $\chi^2 = 9{\cdot}05$ was significant at the ·05 level. So we have a numerical description of the association between the number of fields abandoned on north or south facing slopes, and we may also assume with 95% confidence that, provided our sample is random, it is an association which exists in the 'parent' population of fields abandoned on north and south slopes in that area.

Exercises

1. The slope profiles across one section of a small north-south running valley are measured at seven randomly chosen locations, and the mean slope from stream edge to watershed calculated for each side. The information obtained is as follows:

Profile number	Mean slope angle	
	North side	South side
1	4·5	4·5
2	3·5	5
3	3	3·5
4	6	7·5
5	5·5	4
6	6·5	6·5
7	7	8

Using Kendall's correlation coefficient τ state your hypotheses and show whether there is any significant association between the angles of slope on the north and south sides of the valley, and, if there is, at what level.

2. An investigation is carried out on a steep uncultivated slope below a relatively flat area of cultivated farmland. The bedrock on which the slope has formed is homogeneous and tends to produce an acid type of soil. The

vegetation, however, seems to indicate an increase in alkalinity downslope, which may be the result of surface run-off from the cultivated land above on which the farmer had used lime. Soil samples are therefore taken and tested to establish their pH value at intervals of 100 metres from the end of cultivation to the bottom of the slope. The results of the tests are as follows:

Distance from end of cultivation	100	200	300	400	500	600	700	800
pH value	5·7	5·5	6·2	5·5	6·4	6·5	6·0	6·6

State your hypotheses and test whether there is an association between distance from the cultivated edge and the accumulation of alkaline material in the soil. Is your answer significant? If so, is it at (a) the ·05 level, or (b) the ·01 level?

3. A series of soil samples are taken randomly down a slope profile. The angle of the profile at each sampling point is recorded. The samples are first weighed. Each sample is heated to burn off the organic content of the soil, and then re-weighed. The weight of the organic fraction that has been removed is then expressed as a percentage of the total weight of the sample. The data derived are as follows:

Sample	A	B	C	D	E	F	G	H	I
Slope angle in degrees	1·0	0·5	1·2	2·1	1·5	1·3	3·0	4·0	5·4
Percentage organic content	43	34	38	29	47	55	26	20	33

Sample	J	K	L	M	N	O	P	Q	R
Slope angle in degrees	4·9	4·7	6·0	6·8	7·3	10	8·5	13	17
Percentage organic content	15	24	18	11	7	5	8	30	9

(a) To determine whether there is any relationship between slope angle and the organic content of the soil, state your hypotheses and calculate Kendall's rank correlation coefficient τ.

(b) (i) At what level of significance is τ?

(ii) What is the probability that the apparent relationship between slope angle and organic soil content may be due to random variations?

4. The table below shows the per capita G.N.P. (Gross National Product) in U.S. dollars, and the Infant Mortality Rate (infant deaths for every 1000 live births), for twenty-six randomly selected countries.

$\bar{x} = 470.27$
$s_x = 485.8$
$\bar{y} = 77.2$
$s_y = 44.8$

	Per capita G.N.P.	Inf. mort.		Per capita G.N.P.	Inf. mort.
Canada	1667	31·3	Iraq	195	34·7
Switzerland	1229	26·5	Peru	140	94·8
New Zealand	1249	24·5	Ghana	135	113·0
Sweden	1165	17·4	Egypt	133	140·4
France	1046	38·6	Tunisia	131	173·0
U.K.	998	25·9	Indonesia	127	100·4
Norway	969	20·6	Ceylon	122	71·5
Venezuela	762	69·9	Paraguay	108	70·6
Austria	532	45·6	India	72	99·9
Argentina	374	62·3	Nigeria	70	82·5
Colombia	330	104·2	Bolivia	66	97·2
Malaya	298	73·2	Pakistan	56	100·9
Phillipines	201	110·9	Burma	52	177·6

Source: Atlas of Economic Development, ed. N. Ginsberg (Univ. of Chicago Press, 1966)

Using Spearman's rank correlation (r_s), test whether there appears to be an association between per capita G.N.P. and infant mortality. State your hypotheses and the level of significance of your result.

$\sigma_x = 476$

$r = -0.739$ $\sigma_y = 43.9$

$\sum xy = 541014.5$

$\sum x = 12227$

$\sum y = 2006.7$

$\sum x^2 = 11650943$

$\sum y^2 = 205094.03$ $n = 26$

Chapter 7

Simple regression

Introduction

We have already seen that if the values representing two variables are perfectly correlated and plotted on a graph, a straight line can be drawn joining up all the points (Fig. 7.1A). Perfect correlation rarely occurs however. More often, when a strong association exists between two variables, which results in a significant degree of correlation, the plot on the graph resembles Fig. 7.1B. We can no longer draw a straight line through all the points, but we can draw it by eye through the middle, so that the points are distributed as nearly as possible symmetrically on either side of the line through the long axis of the group. This is a 'best fit' line, positioned so that the distance between all the points and the line is at a minimum. In mathematical terms it is a **regression line.** The regression line is a kind of mathematical model, a graphical expression of the relationship between two variables.

Regression analysis can become very complicated indeed, and it can serve many purposes. This chapter is restricted to the simpler techniques. One use of the regression line is to predict, using a limited amount of data, the value of one variable given any corresponding value of the other variable. Probably more generally important at sixth form level is its use to establish **residual values.** These are **the distances of each point from the line,** drawn in Fig. 7.1C. In establishing a residual value we are showing by how much any point differs from the general relationship expected. Large values generally require explanation. More important, increasing use is being made of residuals to form a basis for mapping areal similarities and differences.

Unless there is a significant correlation between two variables, a regression line is of no value. Obviously, if the degree of association measured has a high probability of being a chance relationship a residual value has an equally high probability of being due to chance, and no secure explanation is possible. Even if a highly significant correlation is obtained, *this does not necessarily imply a causal relationship.* Whether changes in one variable are *causing* changes in the other must be decided by the judgment of the observer.

Similarly, when a causal relationship is known to exist, a different value of x may cause a change in y, but the reverse is not necessarily true. Fig. 7.2 is an obvious example, showing the increase in temperature range with latitude. In a situation like this the variable *causing* the change is described

Fig. 7.1 *A. Perfect positive correlation of y and x.*
B. Significant correlation between y and x. The regression line ab summarizes the relationship of all values of x and y.
C. Significant correlation between y and x. The length of the line drawn between each point and the regression line is 'residual' for y at that point

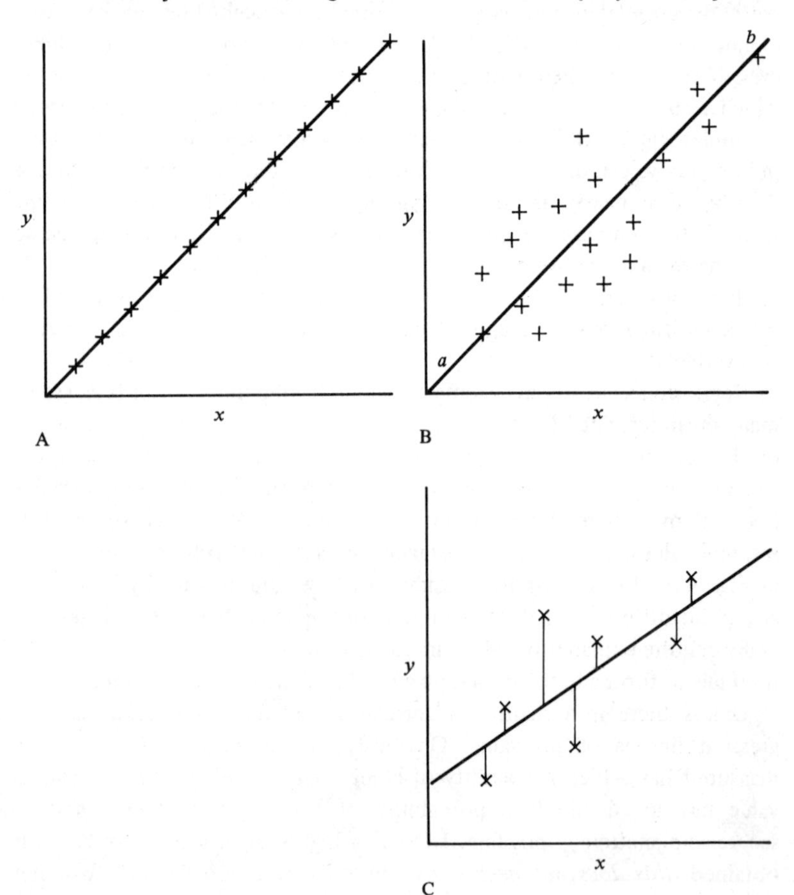

as **independent** and the other, in which the change takes place, is the **dependent** variable. The independent variable is normally shown along the x axis of the graph.

Table 7.1 *Latitude and temperature range in °C for the 30 stations used as data for Figs. 7.2 and 7.3*

	Lat. x	Range y		Lat. x	Range y
Manaos	3S	1·7	Montevideo	35S	11·7
Zanzibar	6S	3·3	Spitzbergen	78N	24·4
Honolulu	21N	4·4	McMurdo Sound	78S	22·2
Durban	30S	7·2	Cairo	30N	15·0
Caracas	10N	2·2	Antofagasta	24S	6·7
Ocean Is.	1S	0·2	Wellington	41S	8·3
Bombay	19N	5·6	Fuchow	26N	17·8
Benares	25N	17·2	New Orleans	30N	15·6
Algiers	37N	13·9	Tokyo	36N	22·8
Valparaiso	33S	6·7	Peking	37N	30·6
Sydney	34S	10·6	Aberdeen	57N	10·0
Boston	42N	25·0	Leningrad	60N	25·0
Lisbon	39N	11·1	Montreal	46N	30·9
Capetown	34S	8·3	Jacobabad	28N	22·8
Copenhagen	56N	16·7	Astrakhan	46N	32·2

Mean latitude value $\bar{x} = 34{\cdot}73$. Mean temperature range value $\bar{y} = 14{\cdot}34$.
Standard deviation of x, $s_x = 18{\cdot}69$.
Standard deviation of y, $s_y = 9{\cdot}31$.
The Product Moment correlation coefficient between x and y, $r_{xy} = +0{\cdot}64$.

The semi-average line

Fitting a regression line by eye is a very inexact process especially when a large number of points is involved. For example, to guess the position of the line in Fig. 7.2 would provide a very dubious result. There is a method using semi-averages which, while still not very precise, is much better than guesswork, and quick and easy to calculate.

First calculate the mean of all the values of x and y–in this case $\bar{x} = 34{\cdot}73$, and $\bar{y} = 14{\cdot}34$–and mark the position on the graph where these coordinates intersect. Then calculate the mean of values of x and y *below* 34·73 and 14·34 respectively, in this case 22 and 5 approximately, and mark on the graph the point where the coordinates intersect. Similarly calculate the means for values of x and y *above* 34·73 and 14·34 (49 and 20 approximately), and again mark the intersection of these coordinates on the graph.

The three points established are very nearly in line. The regression line is now drawn by eye as a best fit line *for these three points*, a relatively simple matter compared with trying to do it for all the other 30 points on the graph.

Fig. 7.2 *Mean annual temperature range of 30 stations. 19 are north, and 11 are south, of the equator. The regression line is calculated using the semi-average method. The coordinates of point A are the means of* all *the values of x and y respectively.* $\bar{x} = 34{\cdot}73$, $\bar{y} = 14{\cdot}34$. *The coordinates of B are the means of the values of x and y* below *34·73 and 14·34, and of C are the means of values of x and y* above *34·73 and 14·34*

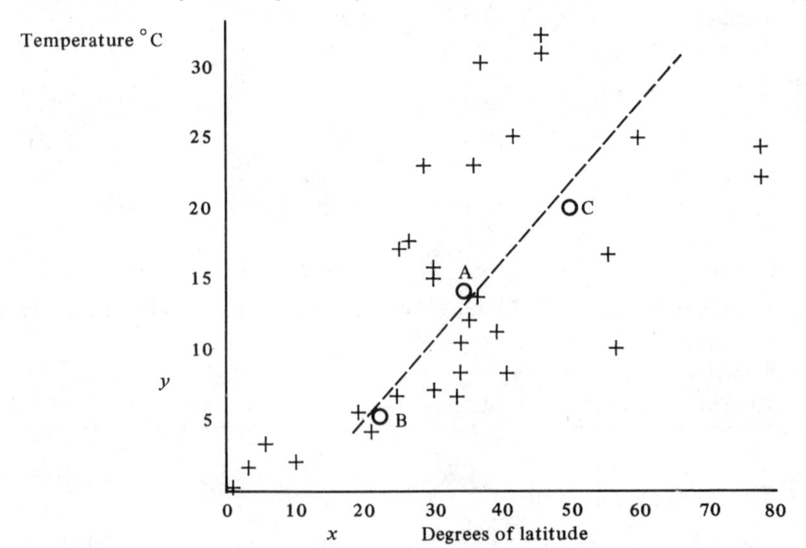

The regression line thus established represents a summary of the relationship between latitude and temperature range based on the evidence of the 30 stations used. From it an approximation can be made of the expected temperature range likely to be found for any given latitude. (A glance at Fig. 7.2 shows how much of an approximation it is.) Two immediate uses can be made of the information the graph discloses. Firstly, the *explanation* of some of the residuals provides an interesting climatological exercise.

Some of them are large, and positive in that they are *above* the line, representing stations with a range *greater* than expected. Similarly, large residuals also exist *below* the line, representing stations where the range is considerably *less* than expected.

Secondly, from a regression graph like this it would be possible to determine the residuals of temperature range for a very large number of places on a world basis. Isolines might then be interpolated to form a map which distinguished between areas *according to the amount they differed from the generally observed association.*

This method of generating information to map areal or locational differences is capable of being adapted to many situations, providing paired data are involved; for example, the relationship mentioned in Chapter 6 between per capita income and the rate of unemployment, or possibly between crop yield and growing season rainfall.

The least squares regression line

The semi-average method of drawing a regression line serves as a quick and simple method of achieving a result and is often justified (especially where no electronic calculator is available) in terms of effort saved, but it is only an approximation of the true 'best fit' line. This is achieved when the sum of the squares of the residuals (i.e. the squares of the distances) of all points from the line is at a minimum. The method of calculating such a line is time-consuming, involving the use of the standard deviation for each set of data, and the Product Moment correlation coefficient 'r'. (The method of obtaining r was not given in Chapter 6, partly because for its use to be valid the restrictive assumption must be made that the population from which the sample is taken is approximately normally distributed, and partly because of the labour involved. But r is essential if the least squares method of regression analysis is to be used, and the calculation of r, for those interested, is briefly explained in Appendix 8a. The explanation of the calculation of r is a memory aid only. It is not intended for those who are meeting regression analysis for the first time, and for whom a full and detailed explanation is required. Such an explanation will be found in *Quantitative Techniques in Geography: An Introduction* by Robert Hammond and Patrick McCullagh [O.U.P., 1974]. An approximately normal distribution may be assumed for the population from which the climatic data contained in Table 7.1 were drawn).

The variables for which the regression line is here calculated are essentially different, in that mean temperature range is considerably affected by latitude, whereas the reverse is manifestly absurd. The same situation might arise if we plotted the abundance of a particular plant species with altitude. The plant species is affected by the altitude, which remains quite independent of any type of vegetation. In both these situations we have an independent and a dependent variable. In the case of temperature range it is quite clear that this depends to a large extent on latitude, and that latitude is the independent factor which varies consistently from the equator to the pole. The presence of a dependent and an independent variable is a common situation. When it occurs it is customary to place the dependent variable on the y axis of the graph, and the independent variable on the x axis. (There may also be occasions when both variables, although they correlate significantly, are essentially independent. In which case the relationship is less easy to define and is generally through a third variable. Interpretation is more complex and this situation, although calculated in exactly the same way, is not considered here.)

To calculate the regression line giving the smallest sum of squares (i.e., squares of the residual distances) of y on x, we have to calculate the mean and standard deviation of both variables, and the association between them expressed by the correlation coefficient r. The means of each variable are

Fig. 7.3 *OP is the best fit regression line of y on x. QR is the best fit regression line of x on y. The pecked line is taken from Fig. 7.2 showing the line drawn by the semi-average method*

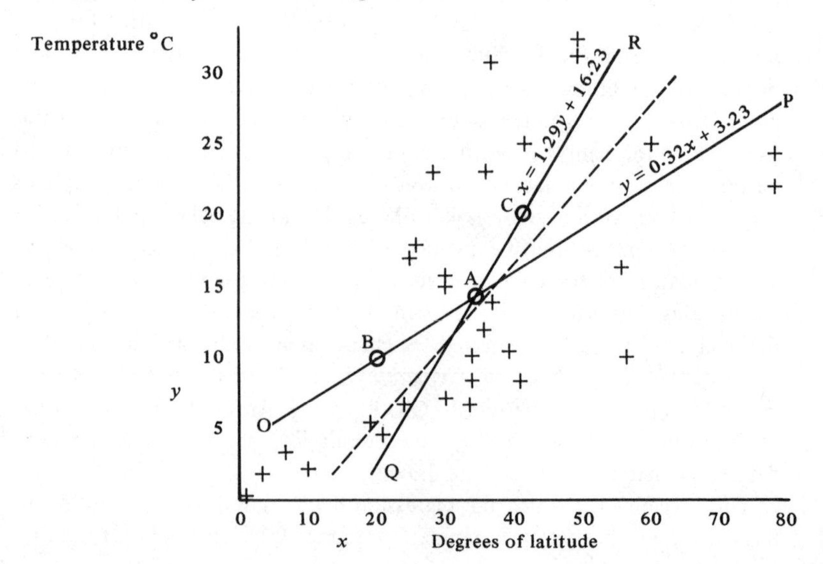

then used as coordinates to fix one point through which the regression line will pass (point A on Fig. 7.3). Another point to determine the angle of the line may then be found from the equation:

$$y - \bar{y} = r \frac{s_y}{s_x} (x - \bar{x})$$

Formula 7.1a

where $\bar{y}$ and $\bar{x}$ are the means of the values represented by y and x, r is the Product Moment correlation coefficient of y and x, and s_y, s_x, are the standard deviations of y and x.

The use of Formula 7.1a to establish a second point for the regression line of y on x is made clear from the values given in Table 7.1. First we have to substitute the required values in Formula 7.1a:

$$y - \bar{y} = r \frac{s_y}{s_x} (x - \bar{x})$$

From Table 7.1 we obtain the following values:

$$\bar{y} = 14{\cdot}34, \bar{x} = 34{\cdot}73,\ s_y = 9{\cdot}31,\ s_x = 18{\cdot}69, r_{xy} = +0{\cdot}64.$$

Therefore,

$$\begin{aligned} y - 14{\cdot}34 &= 0{\cdot}64 \times \frac{9{\cdot}31}{18{\cdot}69} \times (x - 34{\cdot}73) \\ &= 0{\cdot}32x - 11{\cdot}11 \end{aligned}$$

and

$$\begin{aligned} y &= 0{\cdot}32x - 11{\cdot}11 + 14{\cdot}34 \\ &= 0{\cdot}32x + 3{\cdot}23 \end{aligned}$$

We are now able to substitute any value of x we choose and can obtain the corresponding value for y. Let us take $x = 20$ (i.e. 20 on the x-axis) and find the corresponding value for y (on the y-axis) to give us a second point on the regression line.

If $x = 20$:

$$\begin{aligned} y &= (0{\cdot}32 \times 20) + 3{\cdot}23 \\ &= 6{\cdot}4 + 3{\cdot}23 \\ &= 9{\cdot}63 \end{aligned}$$

Thus the coordinates $x = 20$ and $y = 9{\cdot}63$ provide us with the second point (B) for the line drawn in Fig. 7.3. We now have OP, the regression line which minimizes the sum of the squares of all the residuals of y.

If x and y are perfectly correlated then only one line can be drawn, but if correlation is not perfect there will always be two 'best fit' lines, one of y on x and the other of x on y. The former can be used to predict values of y given x. In our case the approximate temperature range may be given for any given latitude.

In cases when both variables are assumed to be independent, the 'best fit' line, of x on y minimizes the squares of the x (horizontal) residuals, and is found from Formula 7.1b.

$$x - \bar{x} = r \frac{s_x}{s_y} (y - \bar{y})$$

Formula 7.1b

Substituting the values already used above:

$$x - 34{\cdot}73 = 0{\cdot}64 \times \frac{18{\cdot}69}{9{\cdot}31} (y - 14{\cdot}34)$$

$$= 1{\cdot}29y - 18{\cdot}5.$$

Therefore, $x = 1{\cdot}29y - 18{\cdot}5 + 34{\cdot}73$

$$= 1{\cdot}29y + 16{\cdot}23$$

If $y = 20$

$$x = (1{\cdot}29 \times 20) + 16{\cdot}23$$

$$= 25{\cdot}8 + 16{\cdot}23$$

$$= 42{\cdot}03$$

The coordinates $y = 20$ and $x = 42{\cdot}03$ now provide us with the second point (C) for QR in Fig. 7.3, which is the regression line which minimizes the sum of the squares of all the residuals of x. QR would not be used unless both variables could be considered independent. In this example it shows an approximation of latitude for a given temperature range, whereas the question we are interested in is a summary of the approximate temperature range in a given latitude. But OP and QR have both been drawn in Fig. 7.3 in order to show that there will often be a considerable difference in slope between the two lines. This is because $r = +0{\cdot}64$ is not very close to 1, though significant at more than the ·01 level. The nearer the correlation coefficient is to 1 the smaller will be the angle between the lines.

The result of the semi-averages method is taken from Fig. 7.2 and superimposed on Fig. 7.3 in the form of a pecked line. It indicates a reasonable compromise between the regression lines calculated by the least squares method.

A regression line, unless the correlation coefficient is very high indeed, is not very useful for making predictions, because the approximation is so general. But at this level of study it has two great values:

1. It generates residuals which can be used to establish locational or areal differences, and to form the basis for mapping them.
2. It is a kind of mathematical model of the general association between two variables (in this case latitude and temperature range).

Exercises

EXERCISES

1. Using the figures given in Chapter 6, Exercise 3, plot the values as a dispersion graph, with slope angle as the independent variable. Draw the regression line showing the relationship between the two variables using the semi-average method.
2. (a) Plot the values given in Chapter 6, Exercise 4, in the form of a dispersion graph, with per capita G.N.P. as the independent variable.
   (b) (i) Calculate the Product Moment correlation coefficient, (r), and determine the level of significance.
   (ii) Calculate the regression line for y (dependent) on x (independent).
   (c) Draw the regression line on your graph. Compare it with a regression line drawn using the semi-average method.
   (d) Compare r with r_s, obtained in the previous chapter. Why should they be slightly different?

Appendix 1

Probabilities associated with values of z in a normal distribution

Note: If we wish to establish a probability known to be *above* $+z$ the value shown in Column B must be halved. The same is the case when the probability is known to be *below* $-z$.

If it is not known whether the z value is plus or minus, then probability is shown in Column B.

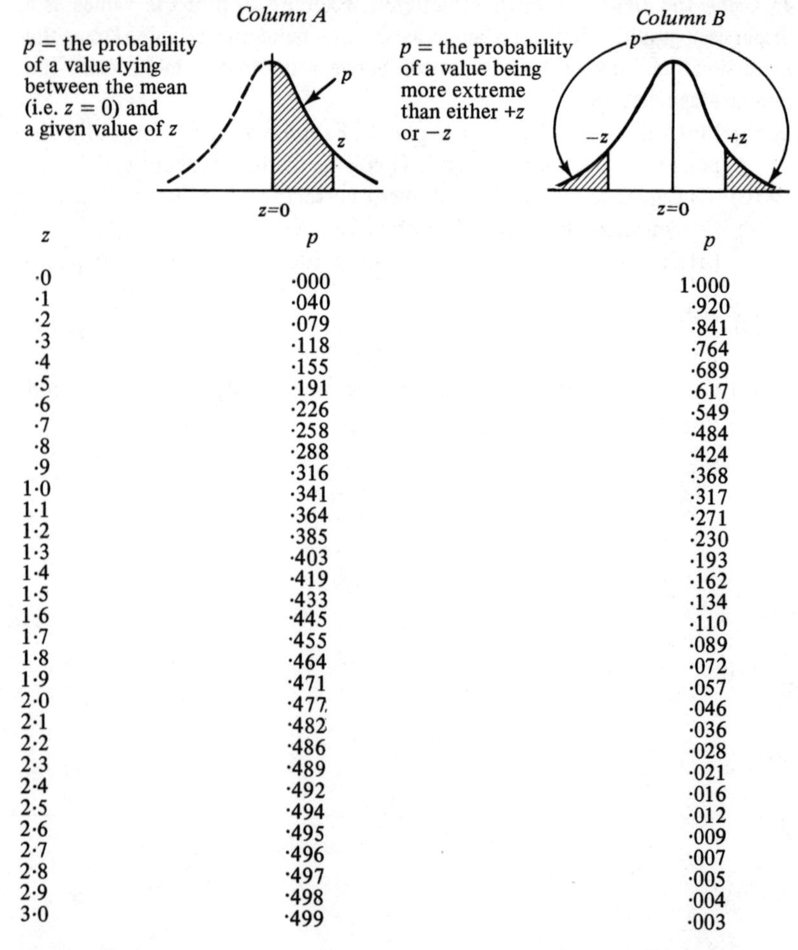

z	Column A p	Column B p
·0	·000	1·000
·1	·040	·920
·2	·079	·841
·3	·118	·764
·4	·155	·689
·5	·191	·617
·6	·226	·549
·7	·258	·484
·8	·288	·424
·9	·316	·368
1·0	·341	·317
1·1	·364	·271
1·2	·385	·230
1·3	·403	·193
1·4	·419	·162
1·5	·433	·134
1·6	·445	·110
1·7	·455	·089
1·8	·464	·072
1·9	·471	·057
2·0	·477	·046
2·1	·482	·036
2·2	·486	·028
2·3	·489	·021
2·4	·492	·016
2·5	·494	·012
2·6	·495	·009
2·7	·496	·007
2·8	·497	·005
2·9	·498	·004
3·0	·499	·003

Source: Lindlay, D.V., and Miller, J.C.P., *Cambridge Elementary Statistical Tables* (Cambridge, 1953)

Appendix 2

Critical values of U in the Mann Whitney U-test for sample sizes $n_2 = 3$ to $n_2 = 20$

Note: These are exact probabilities.

$n_2 = 3$

U	n_1		
	1	2	3
0	·250	·100	·050
1	·500	·200	·100
2	·750	·400	·200
3		·600	·350
4			·500
5			·650

$n_2 = 4$

U	n_1			
	1	2	3	4
0	·200	·067	·028	·014
1	·400	·133	·057	·029
2	·600	·267	·114	·057
3		·400	·200	·100
4		·600	·314	·171
5			·429	·243
6			·571	·343
7				·443
8				·557

$n_2 = 5$

U	n_1				
	1	2	3	4	5
0	·167	·047	·018	·008	·004
1	·333	·095	·036	·016	·008
2	·500	·190	·071	·032	·016
3	·667	·286	·125	·056	·028
4		·429	·196	·095	·048
5		·571	·286	·143	·075
6			·393	·206	·111
7			·500	·278	·155
8			·607	·365	·210
9				·452	·274
10				·548	·345
11					·421
12					·500
13					·579

$n_2 = 6$

U	n_1					
	1	2	3	4	5	6
0	·143	·036	·012	·005	·002	·001
1	·286	·071	·024	·010	·004	·002
2	·428	·143	·048	·019	·009	·004
3	·571	·214	·083	·033	·015	·008
4		·321	·131	·057	·026	·013
5		·429	·190	·086	·041	·021
6		·571	·274	·129	·063	·032
7			·357	·176	·089	·047
8			·452	·238	·123	·066
9			·548	·305	·165	·090
10				·381	·214	·120
11				·457	·268	·155
12				·545	·331	·197
13					·396	·242
14					·465	·294
15					·535	·350
16						·409
17						·469
18						·531

$n_2 = 7$

U	n_1 = 1	2	3	4	5	6	7
0	·125	·028	·008	·003	·001	·001	·000
1	·250	·056	·017	·006	·003	·001	·001
2	·375	·111	·033	·012	·005	·002	·001
3	·500	·167	·058	·021	·009	·004	·002
4	·625	·250	·092	·036	·015	·007	·003
5		·333	·133	·055	·024	·011	·006
6		·444	·192	·082	·037	·017	·009
7		·556	·258	·115	·053	·026	·013
8			·333	·158	·074	·037	·019
9			·417	·206	·101	·051	·027
10			·500	·264	·134	·069	·036
11			·583	·324	·172	·090	·049
12				·394	·216	·117	·064
13				·464	·265	·147	·082
14				·538	·319	·183	·104
15					·378	·223	·130
16					·438	·267	·150
17					·500	·314	·191
18					·562	·365	·228
19						·418	·267
20						·473	·310
21						·527	·355
22							·402
23							·451
24							·500
25							·549

$n_2 = 8$

U	n_1 = 1	2	3	4	5	6	7	8
0	·111	·022	·006	·002	·001	·000	·000	·000
1	·222	·044	·012	·004	·002	·001	·000	·000
2	·333	·089	·024	·008	·003	·001	·001	·000
3	·444	·133	·042	·014	·005	·002	·001	·001
4	·556	·200	·067	·024	·009	·004	·002	·001
5		·267	·097	·036	·015	·006	·003	·001
6		·356	·139	·055	·023	·010	·005	·002
7		·444	·188	·077	·033	·015	·007	·003
8		·556	·248	·107	·047	·021	·010	·005
9			·315	·141	·064	·030	·014	·007
10			·387	·184	·085	·041	·020	·010
11			·461	·230	·111	·054	·027	·014
12			·539	·285	·142	·071	·036	·019
13				·341	·177	·091	·047	·025
14				·404	·217	·114	·060	·032
15				·467	·262	·141	·076	·041
16				·533	·311	·172	·095	·052
17					·362	·207	·116	·065
18					·416	·245	·140	·080
19					·472	·286	·168	·097
20					·528	·331	·198	·117
21						·377	·232	·139
22						·426	·268	·164
23						·475	·306	·191
24						·525	·347	·221
25							·389	·253
26							·433	·287
27							·478	·323
28							·522	·360
29								·399
30								·439
31								·480
32								·520

Source: Siegel, S., *Non-Parametric Statistics for the Behavioral Sciences* (McGraw-Hill, 1956) after Mann, H.B., and Whitney, D.R., 'On a test of whether one or two random variables is stochastically larger than the other', *Annals of Mathematical Statistics* (1947), Vol. 18, pp.52–54

Critical values of U for $\alpha = \cdot 02$

n_1	n_2											
	9	10	11	12	13	14	15	16	17	18	19	20
1												
2					0	0	0	0	0	0	1	1
3	1	1	1	2	2	2	3	3	4	4	4	5
4	3	3	4	5	5	6	7	7	8	9	9	10
5	5	6	7	8	9	10	11	12	13	14	15	16
6	7	8	9	11	12	13	15	16	18	19	20	22
7	9	11	12	14	16	17	19	21	23	24	26	28
8	11	13	15	17	20	22	24	26	28	30	32	34
9	14	16	18	21	23	26	28	31	33	36	38	40
10	16	19	22	24	27	30	33	36	38	41	44	47
11	18	22	25	28	31	34	37	41	44	47	50	53
12	21	24	28	31	35	38	42	46	49	53	56	60
13	23	27	31	35	39	43	47	51	55	59	63	67
14	26	30	34	38	43	47	51	56	60	65	69	73
15	28	33	37	42	47	51	56	61	66	70	75	80
16	31	36	41	46	51	56	61	66	71	76	82	87
17	33	38	44	49	55	60	66	71	77	82	88	93
18	36	41	47	53	59	65	70	76	82	88	94	100
19	38	44	50	56	63	69	75	82	88	94	101	107
20	40	47	53	60	67	73	80	87	93	100	107	114

Critical values of U for $\alpha = \cdot 05$

n_1	n_2											
	9	10	11	12	13	14	15	16	17	18	19	20
1												
2	0	0	0	1	1	1	1	1	2	2	2	2
3	2	3	3	4	4	5	5	6	6	7	7	8
4	4	5	6	7	8	9	10	11	11	12	13	13
5	7	8	9	11	12	13	14	15	17	18	19	20
6	10	11	13	14	16	17	19	21	22	24	25	27
7	12	14	16	18	20	22	24	26	28	30	32	34
8	15	17	19	22	24	26	29	31	34	36	38	41
9	17	20	23	26	28	31	34	37	39	42	45	48
10	20	23	26	29	33	36	39	42	45	48	52	55
11	23	26	30	33	37	40	44	47	51	55	58	62
12	26	29	33	37	41	45	49	53	57	61	65	69
13	28	33	37	41	45	50	54	59	63	67	72	76
14	31	36	40	45	50	55	59	64	67	74	78	83
15	34	39	44	49	54	59	64	70	75	80	85	90
16	37	42	47	53	59	64	70	75	81	86	92	98
17	39	45	51	57	63	67	75	81	87	93	99	105
18	42	48	55	61	67	74	80	86	93	99	106	112
19	45	52	58	65	72	78	85	92	99	106	113	119
20	48	55	62	69	76	83	90	98	105	112	119	127

Source: Siegel, S., *Non-Parametric Statistics for the Behavioral Sciences* (McGraw-Hill, 1956) after Auble, D., *Bulletin of the Institute of Educational Research of India University* (1953), Vol. 1, No. 2

Appendix 3

Critical values of chi-squared

df	Levels of significance	
	·05	·01
1	3·84	6·64
2	5·99	9·21
3	7·82	11·34
4	9·49	13·28
5	11·07	15·09
6	12·59	16·81
7	14·07	18·48
8	15·51	20·09
9	16·92	21·67
10	18·31	23·21
11	19·68	24·72
12	21·03	26·22
13	22·36	27·69
14	23·68	29·14
15	25·00	30·58
16	26·30	32·00
17	27·59	33·41
18	28·87	34·80
19	30·14	36·19
20	31·41	37·57
21	32·67	38·93
22	33·92	40·29
23	35·17	41·64
24	36·42	42·98
25	37·65	44·31
26	38·88	45·64
27	40·11	46·96
28	41·34	48·28
29	42·56	49·59
30	43·77	50·89

Source: Siegel, S., *Non-Parametric Statistics for the Behavioral Sciences* (McGraw-Hill, 1956). This table is taken from Table 4 of Fisher, R.A., and Yates, F., *Statistical Tables for Biological, Agricultural and Medical Research*, published by Longman Group Ltd., London, 1963 (previously published by Oliver & Boyd, Edinburgh), by permission of the authors and publishers.

Appendix 4

Critical values of *r* in the runs test

Table A

n_1	n_2: 2	3	4	5	6	7	8	9	10	11	12	13	14	15	16	17	18	19	20
2											2	2	2	2	2	2	2	2	2
3					2	2	2	2	2	2	2	2	2	3	3	3	3	3	3
4				2	2	2	3	3	3	3	3	3	3	3	4	4	4	4	4
5			2	2	3	3	3	3	3	4	4	4	4	4	4	4	5	5	5
6		2	2	3	3	3	3	4	4	4	4	5	5	5	5	5	5	6	6
7		2	2	3	3	3	4	4	5	5	5	5	5	6	6	6	6	6	6
8		2	3	3	3	4	4	5	5	5	6	6	6	6	6	7	7	7	7
9		2	3	3	4	4	5	5	5	6	6	6	7	7	7	7	8	8	8
10		2	3	3	4	5	5	5	6	6	7	7	7	7	8	8	8	8	9
11		2	3	4	4	5	5	6	6	7	7	7	8	8	8	9	9	9	9
12	2	2	3	4	4	5	6	6	7	7	7	8	8	8	9	9	9	10	10
13	2	2	3	4	5	5	6	6	7	7	8	8	9	9	9	10	10	10	10
14	2	2	3	4	5	5	6	7	7	8	8	9	9	9	10	10	10	11	11
15	2	3	3	4	5	6	6	7	7	8	8	9	9	10	10	11	11	11	12
16	2	3	4	4	5	6	6	7	8	8	9	9	10	10	11	11	11	12	12
17	2	3	4	4	5	6	7	7	8	9	9	10	10	11	11	11	12	12	13
18	2	3	4	5	5	6	7	8	8	9	9	10	10	11	11	12	12	13	13
19	2	3	4	5	6	6	7	8	8	9	10	10	11	11	12	12	13	13	13
20	2	3	4	5	6	6	7	8	9	9	10	10	11	12	12	13	13	13	14

Values of *r* which are equal to or smaller than those shown in Table A, or equal to or larger than those shown in Table B, may be considered significant at the ·05 level.

Table B

n_1	n_2 2	3	4	5	6	7	8	9	10	11	12	13	14	15	16	17	18	19	20
2																			
3																			
4				9	9														
5			9	10	10	11	11												
6			9	10	11	12	12	13	13	13	13								
7				11	12	13	13	14	14	14	14	15	15	15					
8				11	12	13	14	14	15	15	16	16	16	16	17	17	17	17	17
9					13	14	14	15	16	16	16	17	17	18	18	18	18	18	18
10					13	14	15	16	16	17	17	18	18	18	19	19	19	20	20
11					13	14	15	16	17	17	18	19	19	19	20	20	20	21	21
12					13	14	16	16	17	18	19	19	20	20	21	21	21	22	22
13						15	16	17	18	19	19	20	20	21	21	22	22	23	23
14						15	16	17	18	19	20	20	21	22	22	23	23	23	24
15						15	16	18	18	19	20	21	22	22	23	23	24	24	25
16							17	18	19	20	21	21	22	23	23	24	25	25	25
17							17	18	19	20	21	22	23	23	24	25	25	26	26
18							17	18	19	20	21	22	23	24	25	25	26	26	27
19							17	18	20	21	22	23	23	24	25	26	26	27	27
20							17	18	20	21	22	23	24	25	25	26	27	27	28

Source: Swed, Frieda S., and Eisenhart, C., 'Tables for testing randomness of grouping in a sequence of alternatives', *Annals of Mathematical Statistics* (1943) Vol. 14, pp.83–86.

Appendix 5

Critical values of T in the Wilcoxon rank test.

N	Levels of significance	
	·05	·01
6	0	—
7	2	—
8	4	0
9	6	2
10	8	3
11	11	5
12	14	7
13	17	10
14	21	13
15	25	16
16	30	20
17	35	23
18	40	28
19	46	32
20	52	38
21	59	43
22	66	49
23	73	55
24	81	61
25	89	68

Source: Siegel, S., *Non-Parametric Statistics for the Behavioral Sciences* (McGraw-Hill, 1956) after Wilcoxon, F., *Some Rapid Approximate Statistical Procedures* (American Cyanamid Company, 1949)

Appendix 6

Probabilities associated with calculated values of S as large as those shown below for Kendall's rank correlation coefficient

Note: The larger the S value, the more significant the result for a given value of N.

S	Values of N				S	Values of N		
	4	5	8	9		6	7	10
6	·042							
8		·042			9	·068		
10		·0083			11	·028	·068	
12					13	·0083	·035	
14			·054		15		·015	
16			·031	·060	17		·0054	
18			·016	·038	19			·054
20			·0071	·022	21			·036
22				·012	23			·023
24				·0063	25			·014
					27			·0083
					29			·0046
					31			·0023
					33			·0011

Source: Kendall, M.G., *Rank Correlation Methods,* 4th edn. (Charles Griffin and Company, Ltd., 1948) Appendix Table 1, p.173

Appendix 7

Critical values of r_s for Spearman's rank correlation coefficient

N	Levels of significance	
	·05	·01
4	1·000	
5	·900	1·000
6	·829	·943
7	·714	·893
8	·643	·833
9	·600	·783
10	·564	·746
12	·506	·712
14	·456	·645
16	·425	·601
18	·399	·564
20	·377	·534
22	·359	·508
24	·343	·485
26	·329	·465
28	·317	·448
30	·306	·432

Source: Olds, E.G., in *Annals of Mathematical Statistics* (1938), Vol.9, pp.133–48, and (1949),,Vol.20, pp.117–18

Appendix 8a

Calculating the product moment correlation coefficient *r*

The formula for the calculation of r is as follows:

$$r = \frac{1/n \; \Sigma(a - \bar{a})(b - \bar{b})}{\sigma_a \sigma_b}$$

where $\bar{a}$ is the mean of one variable and a the value of each paired variate; $\bar{b}$ is the mean of the other variable and b the value of each paired variate; σ_a is the standard deviation of variable a; σ_b is the standard deviation of variable b; and n is the number of pairs.

The following is an alternative formula, yielding the same result but easier to calculate if a machine is used:

$$r = \frac{\dfrac{\Sigma a.b}{n} - \bar{a}.\bar{b}}{\sigma_a . \sigma_b}$$

The significance of r may be tested from the 't' table shown in Appendix 8b using the formula:

$$t = \frac{r.\sqrt{n-2}}{\sqrt{1-r^2}}$$

Note: This appendix is designed to serve only as a memory aid to those who are already familiar with the calculation of the Product Moment correlation coefficient.

Appendix 8b

Distribution of '*t*' probability

Note: Degrees of freedom (*df*) are the number of pairs less two.

df	·05	·01
1	12·706	63·657
2	4·303	9·925
3	3·182	5·841
4	2·776	4·604
5	2·571	4·032
6	2·447	3·707
7	2·365	3·499
8	2·306	3·355
9	2·262	3·250
10	2·228	3·169
11	2·201	3·106
12	2·179	3·055
13	2·160	3·012
14	2·145	2·977
15	2·131	2·947
16	2·120	2·921
17	2·110	2·898
18	2·101	2·878
19	2·093	2·861
20	2·086	2·845
21	2·080	2·831
22	2·074	2·819
23	2·069	2·807
24	2·064	2·797
25	2·060	2·787
26	2·056	2·779
27	2·052	2·771
28	2·048	2·763
29	2·045	2·756
30	2·042	2·750
40	2·021	2·704
60	2·000	2·660
120	1·980	2·617
∞	1·960	2·576

Source: Backhouse, J.K., *Statistics. An Introduction to Tests of Significance* (Longman, 1967). This table is taken from Table 3 of Fisher, R.A., and Yates, F., *Statistical Tables for Biological, Agricultural and Medical Research*, published by Longman Group Ltd., London, 1963 (previously published by Oliver & Boyd, Edinburgh), by permission of the authors and publishers.

Bibliography

A

Teaching Geography Series (Geographical Association)

No. 6. Newman, R.J.P., *Field work using questionnaires and population data* (1969).

No. 13. Robinson, R.J., *Latin America's economic situation: use of the rank correlation coefficient* (1970).

No. 14. Morrison, A., *Traffic study as quantitative field work* (1970).

School Mathematics Project, G.C.E. O Level series (C.U.P., 1969), Books 4 and 5.

Midlands Mathematics Experiment, revised edn. (Harrap, 1970), GCE, Vol.IIC.

B

Abler, R., Adams, J.S., and Gould, P., *Spatial Organizations* (Prentice-Hall, 1971).

Gregory, S., *Statistical Methods and the Geographer*, 2nd edn. (Longman, 1963).

Hammond, R., and McCullagh, P., *Quantitative Techniques in Geography–an Introduction* (Clarendon Press, 1974).

Contents of Science in Geography, books 1, 2, and 3

S.I.G. 1 Developments in geographical method

Brian P. FitzGerald

S.I.G. 2 Data collection

Richard Daugherty

S.I.G. 3 Data description and presentation

Peter Davis

Answers to exercises

Chapter 2

1. Less than 1 in 100.
2. Less than 5 and more than 1 in 100
3. Probability is less than 1 in 100 that abandoned holdings are due to chance variations.
4. (i) 50% (ii) 16% (iii) 16% (iv) 6·7% (v) 4·5%
   (vi) 0·2% (vii) 68% (viii) 65·9% (ix) 38·1% (x) 2·5%
5. (b) About 12.

Chapter 3

2. a. $\bar{x} = 71{\cdot}2 \quad \sigma = 11{\cdot}7$
   b. (i) 33% (ii) 29% (iii) 11% (iv) 20%

Chapter 4

1. (a) Not significant. (b) At the ·02 level.
2. (a) $\bar{x}_A = 7{\cdot}6, \quad \bar{x}_B = 10{\cdot}2$.
   (b) 25% (c) at ·05 level (d) 1·6%
3. At ·05 level.
4. Not significant.
5. Not significant.
6. Monday 26360, Thursday 24968. Difference is significant at the ·05 level.
7. (a) 2·1% (c) between ·05 and ·01 levels.

Chapter 5

1. (a) With 95% 45 ± 3. With 99% ± 4·5.
   (b) With 95% 45 ± 2. With 99% ± 3.
2. (a) With 95% 920 ± 28·5. With 99% 920± 42·7.
   (c) With 95% 920 ± 21·6. With 99% 920 ± 32·4.
3. (a) About 5mm. (b) About 15mm.
4. 8000 ± 720.
5. (a) 225 (b.i) 234 (b.ii) 525.

Chapter 6

1. $\tau = 0{\cdot}62$, significant at ·05 level.
2. At the ·05 level.
3. (a) $\tau = -0{\cdot}66$ (b.i) Higher than ·01 level (b.ii) Less than 0·3%.
4. $r_s = -0{\cdot}742$. Higher than ·01 level.

Chapter 7

2. (b.i) −0·740. Higher than ·01 level.